W9-CDV-973

SUSTAINABLE GARDEN

Layered planting protects the soil and provides homes and forage for wildlife.

SUSTAINABLE GARDEN

Projects, insights and advice for the eco-conscious gardener

MARIAN BOSWALL

Contents

6 Introduction

8 Living

11 A small part of nature
12 How do you want to live?
14 Before you begin
17 Working with nature
18 Planning ahead
20 Clean energy
22 Sustainable tool kit
27 Time to connect and notice
30 Plants by the door

34 Creating space

—

36 Planning the site
38 Boundaries
44 Materials
56 Starting with soil
67 The wonder of worms
68 Water, water
75 Plastic: from love to hate

82 Gathering

—

85 Making memories
86 Choosing furniture

104 Cultivating place

—

107 Plants for purpose and place
108 Sourcing plants
111 Choosing plants
122 Plants for free
133 The vegetable garden
147 Garden in tune with the seasons and the moon
148 Gardening for your gut

154 Conclusion
156 Index
159 Further reading

Introduction

From the moment we wake up each morning, we are making decisions that affect our well-being and our environment. Many of them may not seem to make much impact, but collectively they have the capacity to do great good. As a landscape architect of twenty years, I make design decisions every day, and early on in my career I sometimes wished I'd had better information to help make those choices and to explain them to my clients.

Over time, I have developed an approach that is part practical and part spiritual. I listen to the land and the people that live there, and each time aim to create somewhere that is both beautiful and nurturing, for the people and for all the other beings that share the space, from below ground up to the sky. On a practical level, I have built up a database of methods and materials that I like, drawing on how they look and feel, where they come from, their longevity, and how much good or harm is done in using them.

In this book, I share my knowledge with you by showing how a sustainable garden can be beautiful, kind and pretty easy to make and maintain, and helping you to navigate the little decisions along the way that add up to make a big impact.

There are lots of practical projects and images to orient and inspire you, but most importantly I hope that the way of thinking will resonate. We are each a small part of a beautiful world and joined together we have a huge part to play.

I find gardening is one of the most intimate ways to experience the infinite power of nature and my place within it. When I design a garden I put into practice inspirations, memories and knowledge that we all share at some level and can use to understand and create our gardens.

Walking in the woods, you may have felt the change in light and temperature under different trees, and seen the way the taller ones grow up through shrubs and bushes, with smaller plants and flowers covering the ground below. This is the way plants in a garden would naturally grow, too, and we can use this knowledge to create a self-sustaining ecosystem. Your subconscious mind will have taken in all these patterns, layers, sounds and smells, and on a physical level our bodies respond with endorphins, which make us feel calm and happy. They tell us that being in this natural place is good for us.

You may also have experienced the miracle of a tiny seed germinating in a pot on a windowsill. From it, a stem, leaves and then flowers grow and swell and ripen into fruit. The fruits feed us, the seeds go through us and, helped by our fertile biome, they can grow again – this is the original closed loop economy!

Or you might have witnessed the same miracle in an embryo in a human: from it you and I have grown to take our place on Earth, until our bodies also return to the land, enriching it as we go.

When we garden, we are tapping into this power. Nature knows exactly what to do, and we are a part of the creative cycle. Within it, we can curate just a little bit for a little while: managing our surroundings to be beautiful beyond the surface; choosing what to plant on the piece of the planet that we look after; learning from our results along the way.

The approach in this book may seem new, but it is very old. We have always known how to live with the land, and now a global consciousness is returning, making us aware once again that we need to work in partnership with the Earth. Clever minds are working fast to try to clear up the mess we have made, and we can each contribute. It's a wonderful time to be alive.

Sustainable gardening includes taking time for yourself to connect and relax.

Living

By reading this book you are likely to be one of the lucky people who has more than enough to eat, with a warm home and a safe place to live. But you may also be one of the growing number worried that our perception of prosperity is an illusion, that there is a danger and a deficiency on its way if we don't start living within our means. So, what does 'living within our means' actually look like and how can we do it on a day-to-day basis? What can we do to help, without taking to the streets, and can we influence the way we live as individuals and collectively?

My style of sustainable gardening starts with kindness: to the planet and the other creatures that live here, to other people and, most importantly, to ourselves. If you study plants and fungi you begin to understand that the boundaries between us and other living organisms – between Self and Other – is a question of a few cells and a single thought. Once we consider that each of us has billions of different bacteria and fungi living on our bodies, more in fact than there are people on the planet, we begin to think of ourselves as being less individual and autonomous.

Extend that thought to the garden and it becomes clear how many creatures and life forms we support and are supported by. The garden doesn't need us any more than we need it to survive. Or turn that sentence inside out: living means helping one another.

A small part of nature

The Gaia theory sees the world as a biome in balance, and humankind as just a small part of that whole natural system. So, what does it mean to be a part of nature in the twenty-first century? And how can we marry this with living in a city, travelling to work on the subway and drinking with friends in the pub?

It's easy to measure ourselves against the criteria we are bombarded with when we switch on our computers or mobile phones. We are told to work harder, earn more, travel further and relax more elegantly in a more beautifully photographed place. We can become disillusioned trying to live up to the 'tick box' achievements that make up the paradigm of success in the Western world. Our gardens can offer us sanctuary from this over-curated and judgmental world. In them we can create a place of refuge and safety, filled with healing energy and love, embodied through the way we lay them out and what we plant. And more than that, gardens can offer us perspective and the wisdom to realize that we are a small part of nature ourselves.

Just like plants, we need sunlight, water and food. But to live sustainably we also need companionship, and that is where gardening becomes both a hobby and philosophy. Trees communicate through underground fungal networks, and gardeners are no less chatty. From online advice forums to community gardens and Instagram seed swaps, you will soon meet gardening neighbours and find sources for plenty of advice and fellowship.

Even more refreshing is that we soon find ourselves in another world with a different paradigm, without social media metrics or advertising images to measure our worth and find us wanting. The natural world doesn't much care what we look like or what we earn or how we did at school. Many people speak of a feeling of coming deeply and satisfyingly home when they are in a garden, and sooner or later find that they are quite simply in love – with nature. And once we love something, we want to look after it.

They say that ideas lead to thoughts and thoughts lead to actions, and actions to habits. When we decide to look after nature, it's not long before our habits catch up with our ideas and, until then, an aspiring sustainable gardener is better for nature than no gardener at all.

Gardens can offer us the wisdom to realize that we are a small part of nature ourselves.

How do you want to live?

As we become more aware of environmental degradation, many of us feel overwhelmed with anxiety or guilt, and resigned to the idea that we can't do enough, so why do anything at all. If we do commit to being more sustainable, the need to do the right thing can also add to this angst. However, once we have accepted that we are part of nature and that our job is to be kind to it, we also realize that this means being kind to ourselves. We can learn how to live more sustainably without beating ourselves up along the way because we are not yet experts. We are all different and our approach to sustainability will be different too.

The process of gardening

By deciding how you want to live, I mean choosing the practical steps that are comfortable for you, and then perhaps exploring areas just outside that comfort zone every few months to extend your sustainable repertoire and lighten your footprint.

We can begin by applying the 'reduce reuse recycle' approach to the way we garden. As gardeners we are not immune to advertising or impulse buying. A trip to the garden centre can end up with a car boot full of plastic trays of annual plants that have little benefit to pollinators and that may have also been grown in peat, to give a worst-case scenario, or perhaps new pairs of plastic gloves and badly made bird feeders that are soon broken by squirrels and thrown into landfill.

Many of these buys come from that scattered approach most of us are guilty of – trying to do lots of things in a hurry. A Saturday morning deciding to 'fix' the garden can end in a pile of ill-thought-through pot purchases that may not make it into the ground, let alone survive the season where they are planted.

One of the best things about gardening sustainably is the knowledge that we have time to spend, and that we can grant ourselves this time to enjoy. In a truly sustainable garden, the process of gardening is just as important as the final look, since it is sustaining us as well as the plants. If we garden for ourselves rather than to impress visitors, we can allow our garden to be our playground rather than a show ground.

The process of gardening is as important as the final look.

The tidiness habit

You may be reading this and thinking that you want a sustainable garden but don't want to spend much time gardening. This is also an option, if you are happy to set the framework for nature to do its thing within your tidiness comfort zone. I have a friend whose tidiness habit allowed her to leave just a few weeds in her border each season, until she gradually came to value the bees that visited them more than their presence annoyed her. Perhaps start by reducing impulse buys and instead reuse the plants you have by splitting them or collecting seed, or get together with friends to divide and share plants in the spring or autumn. Even in a manicured clipped and boxed garden you can help by reducing petrol-driven machinery in favour of rakes and shears, creating compost bins to replace plastic refuse bags and replacing artificial fertilizer with the nutrients the composted waste provides.

The journey begins by deciding that you'd prefer to be in the garden rather than just look at it.

Note where the sun rises and sets to frame these special moments.

Before you begin

When planning your garden, it pays to spend some time getting to know it well before you start to lay it out. By working with your site and its natural resources, you can save money, time and heartache, and quickly create a more sustainable and self-supporting mini microclimate. I frequently hear people say that they bought a plant and it died soon after planting, but with a little research this can be avoided, and both plants and people can flourish in a space.

The underlying features

The first question to ask yourself is what would this place be if there were no people? If we take a city like London, for example, we know that it has expanded from being a strategic human settlement in the south-east of an island in the North Sea, positioned on a large river estuary leading out to the sea. This means that it is quite mild, but has wind tunnels, especially near the river. The soil is fertile – historically, much of greater London was used as market gardens that made use of the rich alluvial London clay – and gently rolling slopes create both flat and sloping gardens that can take advantage of the sun at different times of the day. Below the clay the bedrock is mainly sandstone, which drains quite freely, and the water table varies, depending on height and distance from the many rivers and springs, often now hidden underground, that feed the Thames.

Maps are available online that show all of these features, but it is usually enough to look at your garden on a street map, or use the compass on a smartphone, note which way is north and then identify areas of sun and shade, the direction of the prevailing wind, and the plants that are thriving, either in your spot or nearby.

Drawing up a plan

It is a good idea to draw a rough plan of the garden (you can get the layout and north–south orientation from tapping your postcode into Google Earth or StreetFinder) and then mark where the shade and sun falls at different times of day. An east-facing corner will be great for a morning cup of tea or coffee in the sun, but in early spring the sun will defrost some flowers, such as camellias, too quickly and cause browning on their petals. In the Northern Hemisphere, a south-facing wall will absorb heat during the day and provide a lovely sun trap in cooler areas, but it can get too hot unless you add some shading, such as a tree. A north-facing garden will receive very little sunlight and algae or moss may grow on hard surfaces if it is also damp, but these spaces will enjoy the clear light that artists adore and offer a haven for shade-loving plants.

If you look carefully, you can also see what is going on underground. You may see bumps in the lawn where tree roots extend, which means that anything planted there will be competing for water and nutrients, or you might notice bare patches that dry out quickly in summer, possibly indicating the presence of underground drains. Where the level of underground reservoirs of water, known as the water table, is high, thirsty plants will thrive and you could see willow trees or *Carex* sedges sprouting spontaneously. If trees or bushes are leaning to one side it may mean that the site is blustery and they have grown away from the wind, or that they are stretching sideways to allow their leaves to reach the light if their roots are in the shade.

The next step

One of the most important ingredients for any garden is the soil, and we will look at testing and caring for your soil in detail in the next chapter (*see pages 56–65*), since your soil will determine what will grow and thrive, both in your garden and the planet.

Once you have assessed your site's main features, you then need to decide how you would like to use it.

Drought-tolerant plants sit below a gutter used to collect rainwater.

Working with nature

Now you know more about your site, you can explore the resources already available in your garden, which you can use or reuse to minimize what you have to buy or bring in.

Plants rely on sunlight to photosynthesize and grow, so the amount of sun is the number one resource to harness when choosing what to plant and where. By using vertical spaces, as well as horizontal areas on the ground, you can include more plants that can reach up to the sun. You can also plant in layers (*see page 120*), creating tiers of different species that can access the available sunlight. And by including deciduous trees you can plant spring-flowering sun-lovers beneath them to catch the rays before the leaves unfurl, followed in summer by shade-loving species to shelter below their canopies.

The second big resource plants require is water. If you are lucky enough to live in an area of high rainfall, you can collect water from your roof in water butts. It's relatively easy to install a downpipe converter kit to do this. In larger gardens, you can divert rainwater into an underground rainwater harvesting tank, linked to an irrigation system to help establish new plants or supply water during periods of drought. If you want to benefit wildlife as well, another option is to collect rainwater in a pond or a mini rain garden (*see page 72*).

If you plant deciduous trees the leaves will inevitably fall in autumn but rather than raking them all up and seeing them as a chore, you can leave some on beds and borders to rot down and provide a free soil food and conditioner – manna from heaven. Dead plant matter is another free resource when turned into garden compost, as are all the scraps from the kitchen which can be added too, or if you are worried about attracting rodents to the garden, fish and meat scraps can go into a wormery or bokashi bin (*see page 67*).

Another great resource are birds, who will eat bugs such as greenfly and caterpillars. You can co-opt them in by providing plants with berries and seeds, and by leaving the seed heads over winter, thereby helping to keep the garden beautiful and healthy, as well as bringing it to life with birds' movement and song.

Finally, the two human resources you will need to assess are person power and money. By planning a garden to suit your energy and your budget you can feel both fit and abundant.

Planning ahead

Spending time outside is good for us. We know that fresh air and exercise help to keep us physically healthy and recent studies have also shown that being in nature for as little as twenty minutes a day decreases the stress hormone cortisol, while having our hands in the soil increases serotonin, which makes us feeler calmer and more relaxed, and may also increase our immunity to diseases. Even ten minutes spent raking up leaves from a lawn can give you a huge sense of satisfaction and accomplishment – this has been proven in a study of young offenders that showed similar tasks helped to increase their self-esteem.

A garden for enjoyment

There's no better way to let off steam than digging. So, it's a strange paradox to design gardens covered in concrete or artificial grass in order to spend as little time as possible tending them, or to fill air newly cleaned by trees with fumes from leaf blowers. A garden should be a pleasure and a refuge from our working lives, not just an addition to the to-do list, so consider how much time you have and your level of fitness and try to find a balance between effort and relaxation.

First, decide what you enjoy doing, which may take some trial and error. Transplanting seedlings in a glasshouse while listening to the radio could be your idea of bliss, or pruning roses, chasing bees with a camera, digging vegetables, or tending the barbecue. Then add more of whatever you like doing most to your garden plan and reduce the things you don't like, to create greater enjoyment and less guilt or grind. For example, plan border planting to reduce weeding time unless you love weeding and avoid laying large lawns if you don't want to spend your weekends mowing. By replacing half your grass with flowers you'll reduce the time spent cutting it and your annual mower fuel by 50 per cent, and be rewarded with a garden full of colourful blooms and butterflies.

If you have the budget, you could start by planting some trees and hedges for structure. They will add instant impact: giving scale, shelter, seasonality and colour. Trees are also great for biodiversity, offering homes to birds and insects, and require very little care once established – they need just a little water for the first couple of years after planting and then mostly look

Listening to and observing nature are a gardener's best tools.

after themselves. If it's your first garden, perhaps start small with some herbs near the house or vegetables in raised beds and plan to expand as your confidence and repertoire of plants grows, rather than overbuying and regretting it later.

For those with more money than time, it may be worth employing a thoughtful gardener who can garden with you and share tips, while also taking on tasks you are unable to fulfil. Just as not all gardens are sustainable, not all gardeners are either, so check that those you employ have the same ethos – you will be sharing your garden and your gardening journey with them and it may be a long one. It would not be a realistic ambition to play a concerto alone in the first season of learning piano, and gardening is a journey of learning, too, for the mind and the body, as anyone who's spent a day digging will know.

The rhythm of time

The gardening year has a certain rhythm which is easy to get used to in time. Spring and autumn tend to be the busiest and summer and mid-winter more restful. It pays to pin up a sowing and harvesting calendar for a veg garden and to follow a 'gardening year' book for other tasks. And remember that while there will be the best time to do most things, as the late, great gardener Christopher Lloyd said, the second best time is when you can.

Patience and observation are the greatest tools of the sustainable gardener. By noticing what is going on around us we can work with the land effectively and waste less time and energy trying to achieve unrealistic goals. Listening and planning really are time well spent.

Clean energy

Once you've decided how you want to spend your own physical energy in the garden, the next question is what other energy will you need to bring in?

Nearly all energy comes from the sun, whether it's the direct heat of the sun or sunlight turned into carbon by plants through the process of photosynthesis. Plants are then eaten by people, animals or insects, which then consume or are consumed by other creatures in the food chain. The bodies of those plants, insects, animals and people also break down over millions of years to produce fossil fuels. If we look at energy sources in this way, it makes sense to use the most recent and renewable forms to grow plants and look after the garden.

A glasshouse is a very efficient way to maximize solar energy to grow plants. The glass allows all the available sunlight in during the winter and you can prevent plants from scorching in summer by siting it in the shade of a deciduous tree or using blinds. To ensure plants don't freeze in winter, try insulating your glasshouse with recycled bubble plastic, wrapping pots in wool or cardboard or draping plants in hessian on the coldest nights. For additional heat, solar panels are one option or hybrid methods of heating, such as passive solar heating systems, may be more energy-efficient and affordable. The alternative is not to heat your glasshouse at all, propagate hardy plants only and, if you grow tender types such as geraniums (*Pelargonium*), keep them indoors or wrapped up in the glasshouse over winter. You can still stretch the growing season by placing seedlings on a windowsill or sunny spot in the house and transferring them to a glasshouse or cold frame to benefit from the sun after the coldest weather has passed.

Fossil fuels are also used to create synthetic fertilizers and pesticides, which are often sold in plastic bottles. You can replace these with organic soil improvers made with your own compost, and by planting in layers and following a simple integrated pest management system.

If you use a petrol-powered lawnmower, an irrigation system and fertilizer to maintain your lawn, despite its vibrant colour, your grass may be the least 'green' thing in your garden. According to a Swedish study, using a petrol mower for one hour has the same carbon footprint as a 100-mile car trip. The US Environmental Protection Agency (EPA) also found that petrol-powered lawnmowers emit eight times more nitrogen oxides, 3,300 times

Save money by cancelling your gym subscription.

Cut down on mowing to reduce fossil fuel use.

more hydrocarbons, 5,000 times more carbon monoxide and more than twice the carbon dioxide per hour of operation than electric lawnmowers, so if you are wedded to your stripes, consider switching from petrol or diesel to an electrically powered mower.

In addition to the air emissions, the EPA also estimates that, around the world, more fuel is spilled when filling lawnmowers and other garden equipment every year than the oil that was lost in the *Exxon Valdez* disaster in March 1989. That's over 10.8 million gallons of spillage contaminating the soil and the ground water.

Noise pollution is also worth considering. The World Health Organization recommends that general daytime outdoor noise levels should not exceed 55 decibels. The average leaf blower can produce 70–75 decibels at a distance of 15.25m/50ft, which is anti-social for the neighbours as well as harmful to the planet.

Some manufacturers are no longer producing petrol garden tools, and instead are investing in battery-powered lawnmowers, hedge trimmers and leaf blowers. They are worth looking into, but unless you have a huge garden it's also worth considering some simple tools that rely on human power alone. Good tools are a joy to handle and make the job much easier but still save on the gym subscription. When did you last feel that delicious fresh-faced tiredness and satisfaction of completing a good outdoors job? If you don't remember, it could be time to try it.

Sustainable tool kit

Whether you are a first-time gardener or more experienced and reviewing what you have, there are some tools that really help in the garden, but only a few that are really needed. Many can also double up to do more than one job, saving space and resources.

Whatever stage you are at in your gardening journey, it is worth investing in a few good quality tools. Well-made items last a lifetime and tools that give you pleasure to look at and to hold will make gardening a joy. Find them second-hand online or at plant fairs, and choose solid metal tools with wooden handles that can be changed if they wear out or break. Include the following in your kit:

- A trowel and spade
- Gardening gloves
- Watering can and rose head
- Secateurs and shears
- Rake or besom broom
- Recycled plastic pots or buckets
- Lavatory rolls, newspaper and pot-shaped vessels such as jam jars
- Hori hori knife
- Penknife (or scissors) for cuttings
- Pencil for a DIY dibber
- Yoghurt pots for labels

Well-made tools last a lifetime and make gardening a joy.

SMUKIES
Holland

Other useful tools

A sharp trowel can save time and effort and I use a copper one that slices cleanly through the soil and is thought by some people to leave trace elements to help deter slugs and snails and keep soil healthy. A new trowel will come with a long guarantee and while a copper one will dull down in colour if left outside, it can be cleaned easily, sharpened with a whetstone and oiled with camellia oil.

A spade or shovel is useful for shifting quantities of soil or compost. Make sure the spade is not too big or heavy to use comfortably – like repetitions in a workout, it is better to do more light work than strain yourself with heavy tools.

A good pair of secateurs will last a lifetime if you look after them – I use the Japanese Niwaki carbon steel range. Sharp blades require less force to cut, so sharpen and oil your tools regularly. Shears that balance in your hand and glide open and shut make cutting a hedge into a dance – watching a topiarist at work will inspire you to ditch the electric cutter.

A penknife in your pocket for cutting string and wire will save your secateurs and if you are offered one as a present a Japanese hori hori knife doubles up as a fabulous weeding tool.

Bamboo gloves are comfortable, waterproof and fit well enough for close work, although you may need a spare heavier pair for tackling thorns.

A metal watering can with a rose head for seedlings is a must, but an old water bottle with a nozzle cap can be adapted if necessary.

A good rake turns leaf clearing into a mindful moving meditation, while a besom broom works for both lawn and terrace (*see the project to make one on page 80*).

A firm kneeler or plank of wood helps spread your weight and protects the soil structure when the ground is wet.

So many gadgets are unnecessary clutter. A pencil for writing labels can also be used as a dibber to transplant young seedlings. Yoghurt pots that housed seedlings can then be cut up to make labels, while lavatory rolls and newspapers can become pots (*see page 78*) and then compost in turn.

We will look at plastic later, but as a valuable long-lasting resource, it is worth reusing any plastic tools you have for as long as possible, rather than sending them to landfill.

My core kit: bamboo gloves, a kneeler and trug, border fork or spade and a favourite copper trowel.

Time to connect and notice

One of the most valuable resources for a gardener is time. Prompts and advice are useful but gardening consciously is about spending time observing and thinking about what is going on, rather than completing to-do lists and achieving goals. Looking at your garden every day over time is the best way of seeing what is happening and beginning to understand why.

A garden can easily become a chore and if you are stretched for time elsewhere, you may not wish to venture outside at all to avoid the jobs that need doing.

Once we realize that the garden has its own agenda and work with it not against it, our approach changes to one of collaborative enjoyment. If you get up ten minutes earlier and spend time outside before you start your day just watching, you will learn more about sustainable gardening than any book can tell you. While drinking your morning cuppa you might observe which trees the birds are feeding on or the plants the bees are visiting, or even what the slugs have eaten, noting the plants that are growing unharmed and those that are suffering. All this is valuable information even if you do not act on it straightaway. By getting to know a garden in all seasons we become better gardeners, more thoughtful and more in tune with the land. Moments of looking and listening are as vital as the time spent tidying up and giving the garden the look we would like it to have. Notice where the brambles grow, the robin nests and solitary bees burrow, where the frost collects in late spring when it can do the most damage, and where the berries ripen first.

A bench or chair in a sunny spot will help to stretch this time, and bring a notebook or sketch book with you. I like to draw and take notes and photographs, although to be honest I rarely review them, but the act of recording them helps commit them to memory so that next year I am a little wiser.

Another way to stretch time is to meditate in the garden. Whether sweeping leaves or pruning, we soon lose ourselves in a task and learn to leave our worries at the garden gate. The simple meditation spiral project (*see overleaf*) will help you to do just that.

A place to spend time looking and listening is as vital as the time spent gardening.

Project: Make a lawn spiral

A lawn spiral will reduce your mowing by half, which has to be good when research shows that hour-for-hour, petrol-powered lawnmowers produce eleven times as much pollution as a new car. The birds, butterflies and bees will thank you, as will the many unseen insects that live in the sward and provide food for larger species in the food chain.

Less mowing will give you 50 per cent more time for something else, while the spiral will also make your lawn bigger. How? In the same way that if you stretched out your gut, it would fill a tennis court, this spiral path will take many times longer to walk the whole length than to cross the lawn, and becomes a lovely way to remain grounded, connect with nature, and to really notice what is growing in the grass. By simply not cutting the lawn you will be amazed at how many species of grass and flowers appear that were living beneath your mower blades. You may even find some you can forage for the kitchen table or your flask. Self-heal and lemon balm from my spiral are regular additions to my morning tea.

Materials & tools

- Small stakes and string
- Long stick
- Lawnmower

Method

To begin, leave your lawn without cutting until it is twice as long as usual. Then find the centre of the area and place your mower there. You can stretch two diagonal strings across the lawn to find the centre, fixing them to the small stakes.

Tie a stick to the mower that is three times the width of the cutting deck. The stick and mower provide a guide for the long grass and the path, which will be same width as the mower.

Begin mowing in an anti-clockwise direction from the centre outwards, always leaving the width of the mower-attached stick to your left unmown. By going slowly your paths will remain the same width and the spiral will be even. You should be able to see a difference straightaway if your grass was quite long. If you go wrong just stop and go back over it, and if it's a little wonky don't worry, so is life, and the grass will grow!

To maintain the spiral, simply mow the short grass through the summer and leave the longer area to flower. In the spring when the grass starts growing you can cut the longer area short if you want to start again and put the clippings on the compost as usual.

Einhell

Plants by the door

Marie Kondo was right about most clutter, but with such simple joy to be had, pots by the door has to be the place to break the rules. They will greet you when you get home and call you to de-stress and potter in a spare five minutes. The plants can be both beautiful and useful, of course, and pots provide a good home for gifted plants, odd ones that don't fit elsewhere and special types that need a bit of nurturing. They do not have to go with the overall scheme of the garden and offer a great opportunity to experiment with textures and colours to help you to decide what you like.

Start a reclaimed pot collection with any type of container that will hold water but drain gradually. Find an old, galvanized steel washtub or even a bucket online or in a skip and drill holes in the bottom. Fill half with waste polystyrene, plastic pots or bricks to create a good drainage layer and prevent roots from sitting in wet compost. Then fill the top half with garden compost and plant into this. Any container can be adapted, from an old wellington boot to a washing machine drum; this is a spot to get creative and have some fun.

What to plant?

In a large container, try evergreens such as myrtle (*Myrtus communis*), sweet box (*Sarcococca confusa*) or bay (*Laurus nobilis*), which have a lovely scent if you brush past them and can also be cut to bring into the house or used in the kitchen. The birds like the berries, too, and all of these shrubs are tolerant of shade and some air pollution in a city garden. In smaller pots, plant sedums or houseleeks (*Sempervivum*), which are easy to look after and said to bring luck to the house, or some herbs such as thyme or rosemary to snip for your plate or just scent your pocket on your way out. To round off a simple but satisfying collection, climbers can also work well in big pots. Try the late summer-flowering *Clematis* 'Alba Luxurians' or a star jasmine (*Trachelospermum* jasminoides) to add some scent and feed the bees.

A metal trough, chimney pots and tubs filled with pelargoniums make a scented entrance.

1
2
3
4

Project: Make your own herb spiral

Cooking with herbs adds flavour and depth to any ordinary dish, and has the health benefits of fresh nutrients which can be lacking in processed food. To quote the Ancient Greek physician Hippocrates 'Let food be thy medicine, and let medicine be thy food.' It's a great idea to keep herbs as near to the kitchen as possible.

You could start with a pot of basil on your kitchen windowsill to bring some summer to your food, and build up to a collection of herbs with individual watering needs.

This herb spiral is the perfect solution when you have lots of pots and space is at a premium. The helix shape employs every bit of space and uses three dimensions, as do the branches of a tree. Based on a design from nature, it also makes use of all the available sunlight, shade and water, with the plants at the bottom benefitting from water run-off and those above positioned to get their preferred amount of sun.

First, choose plants that you want to eat for their flavours, health benefits or perhaps associations with people or places. Some easy herbs to start with in a temperate climate could be rosemary, sage and chives. Mediterranean plants such as rosemary have thin aromatic leaves with a waxy cuticle designed to hold water, which makes them good for a dry spot at the top of the spiral – and also less attractive to rabbits, which is worth knowing if you live in bunny country. Adjust the size and layout of the spiral to suit your garden and your cooking needs.

Materials & tools

- Large logs, stones or bricks
- Selection of herbs
- Watering can

Method

Start with some large logs, stones or bricks at the bottom and build them up into a spiral shape. Any builders' rubble from a skip will do as long as there is not much cement or any other contaminants left on them. **[1]**

Fill in between the rock or stone layers with soil. You can use subsoil at the bottom, followed by a layer of compost on top. The soil will settle so water it in or add some more later if you find holes forming. **[2]**

Then select your plants for their place in the spiral. Plant large-leaved thirsty plants such as sorrel, parsley and basil at the bottom and the sun-lovers like oregano, thyme and rosemary at the top and on the sunnier side. Parsley and chives can go on the cooler shadier side. You don't have to limit yourself to herbs; add some strawberries, lettuce or salad leaves if you wish. **[3-4]**

These clay pavers and oak benches will outlive the owners but gradually meld back into the land from which they came.

Creating space

Creating a space can be the most magical part of designing a garden. When you look at a blank canvas the potential is limitless. What you do with it comes back to how you want to live, how much you want to look after it and how much time and money you have to invest. To create a sustainable garden, you need to plan for the long term and at the same time consider how much you want to spend today.

If you are renting or intending to move on in a few years' time, it is still worth investing in plants, but you may be able to make do and make over the existing paving or gravel areas. Old mortar can be removed and plants such as thyme sown in the cracks to give uninspiring paving slabs a lift, while adding to your garden's biodiversity and attracting pollinators. By creating gaps between the slabs, you also allow water to permeate and reduce run-off, which helps to protect the soil elsewhere in the garden and also helps to prevent flooding in built-up areas. Gravel is a relatively cheap way to keep feet dry and create areas that can be for both people and plants. Find ways to plant in gravel on page 128.

The way a garden looks and feels is a direct result of your intention for the space, so it is worth deciding at the start if it is primarily a garden for entertaining, meditating or being alone, growing food or allowing children space to play. Of course, with a bit of careful planning it can be all of these things in turn, to all of its users.

Planning the site

The first step when planning your site is to take time to observe. Watch where the sun's first rays find the garden in the morning and note where the last rays linger at sunset. These will be your morning and evening prime spots for yoga, coffee or sundowners. Note that these will change slightly through the year as the Earth orbits the sun (*see also page 15*).

Next, spend time sitting in spots in the garden and feel the different energies. Do you feel calm and peaceful, or excited and exhilarated, or uncomfortable and irritable in each area? Use these findings to make spaces to meditate or potter, or to fill with exuberant planting or to create play areas.

You may also want to improve other areas if they make you feel uncomfortable, such as spaces overlooked by neighbours. You might decide to add screening for privacy or create a social hub to enjoy with them, depending on your mindset. Or are there bright lights at night, loud sounds to mitigate or strong smells to consider? Is the scale comfortable from sky to tall trees to humans, or do you need to plant an interim layer? If your garden is in a dip or shady, it may hold frost or damp and need help to move these on, while sunny areas might be too hot at midday and require screening from the glare. If you are in a wind corridor you may also need shelter or if your garden is on a high water table (*see page 17*) you may need better drainage and ways of holding the water to prevent flooding.

Then think about who else uses your garden: are there certain creatures you want to create a home for or to encourage, or are there marauders you want to dissuade? Does your boundary link with others to provide a wildlife corridor for bats, owls and harvest mice, or if not, could it? Add to your assessment the ways in which you want to spend your time in the garden, in terms of fun and productivity. And finally, make a list of the waste bins, bicycles, compost bins, glasshouse, air source heat pump, water butts, log store, extra children's play areas and other features that will need a home too.

Once you are armed with the site specifics and what you want to do in it you can start to lay out what will go where, and decide how you will move through the spaces and what you will grow: vegetables, flowers, trees, hedges, herbs, climbers, nettles and comfrey. You may want to include partitions or be happy with areas and needs that overlap; nature is very good at sharing space, just don't forget to build in zones for wildlife.

Screening, shading, fruit and wildlife haven are provided by the trees overhead, with a convivial dining spot below.

Boundaries

Defining space within the garden is as important as deciding what to plant in it. It can be easy to fall into the trap of thinking that you need to put down lots of hard landscaping and carve up the garden into 'rooms' with walls and fences, when really it's more a question of deciding how you want to use the space and then creating a different energy for each area. In many gardens it can be just like arranging a room inside your home. Once you've decided where you want to eat, sit, read or potter in a sitting room, you can define each space with something as simple as a chair, table, lamp, plant or a rug. The boundaries in the garden can be defined with equal simplicity. When you know where the sun and shade are, which areas will be more productive and which will be best for rest or play, you can arrange the contents to suit.

Your plants do not need to be permanent or big budget. A strip of wildflowers can separate a smart mown lawn from a play area and be less of a worry if a stray football hits them than if it breaks a row of alliums, for example. A simple wood chip or gravel path can define the space between one part of the garden and the entrance to another area, such as a woodland glade, and can be taken up or moved easily if the garden changes.

A change in planting or materials sends a message that you are entering a new area, and also shows the way when you arrive somewhere new. It can be as simple as planting a tree or shrub where it will act as a welcoming sentinel to guide visitors to the front door or main entrance. In some places these welcome trees can outlive the owners and even the house, so choose a tree that will fit the site when it matures, without needing to be pruned back.

Boundaries can also provide privacy and security, and need to be chosen to do the job effectively, making the people inside feel comfortable without creating a fortress. Whether you want to deter other people, animals or wind, the boundaries you build can be chosen with sustainability in mind.

Space can be defined by high-energy planting, a simple brick path or just a well-positioned chair.

Walls

If you are thinking of creating a solid boundary, you will need to factor in security, cost and aesthetics when choosing between a fence or wall, but with over two billion bricks being used in the UK per year and the fencing market worth over £1 billion a year, the most sustainable option may be to reuse what is there already. If that is not possible, a well-built wall will have a much higher initial cost than a fence or hedge but need minimal on-going maintenance to create a durable and secure barrier. Walls also absorb heat which is emitted as they cool. This means that by planting next to a south- or west-facing wall you can extend your growing season and include more tender species.

For shelter from wind, a permeable fence or hedge will be more effective than a solid barrier because it will filter the wind and slow it down, whereas a wall can exacerbate gusts by creating an eddying effect on the lee side.

For a wall, choose a long-lasting material that will not need replacing in the short term, and opt for materials available locally to minimize transportation. Aesthetically, it is most likely to stand the test of time if it looks beautiful, sits well in the landscape and complements the style of the house. Masonry walls are most frequently used in older properties. However, walls can also be made of timber, metal gabions filled with recycled glass bottles, or even of turf.

The basic structure for a wall is usually a single or double skin brick, or a precast concrete block for strength with a facing of brick or stone. A single brick construction can be used for a wall up to 45cm/18in high. Most perimeter garden walls are two bricks thick and may need piers or buttresses if they are tall. Efficient use of resources may be the single most important strategy when building a sustainable garden, so to reduce the quantity of materials keep walls low and ensure their longevity by including drainage, capping and a damp-proof layer to help protect them from the elements. Or, if you have space and need a tall structure, you could build a single skin, serpentine-shaped 'crinkle-crankle' wall; its curves give it strength without the need for extra support and it offers an unusual and beautiful feature.

Designing for deconstruction rather than demolition prolongs the lifetime of each individual brick, which can be used again if a wall is taken down. For example, cement mortar has a very high carbon footprint and is also hard to remove, damaging bricks on the way, but a lime mortar, which is easier to remove, will allow the wall to be taken down and the bricks used again.

Repair is usually better than replacement if a wall is in need of some TLC, and reclamation yards are a good source of beautiful old bricks. If there are small gaps in old mortar you can leave them for insects and bees to inhabit. Masonry bees will colonize the holes to lay their eggs, while bats also love crevices and rarely do any real damage to a wall. Moss growing on an old wall captures rainwater and provides a good drinking place for bees. It also actively removes air pollutants, including nitrogen dioxide, and is a great plant for shady town and city gardens.

Fences

As the famous American poet Robert Frost said, 'Good fences make good neighbours.'

If a wall is beyond your budget, a fence has the advantage of being more permeable and, as long as it's not a close-board type, can promote neighbourliness. The origin of the word 'fence' comes from the fourteenth-century word 'fens', a term for defence or protection, and even an open-structured fence can help to increase

A mortar-jointed wall, a cedar fence, acers and a solid oak bench provide a layered boundary.

security just as much as a wall because there is less to hide behind. A garden blocked from the road and hidden from others by tall close-board fencing is also a death knell to the sense of community epitomized by a cup of tea and a chat over a neighbours' chin-high fence or hedge.

Fence materials have changed through history and differ by region: in the Bronze Age, people used stone, a solid durable material, for their fences, but the Anglo-Saxons preferred worm fences, which were zigzagging structures of rough wooden rails crossed at an angle that did not require posts driven into the ground and were particularly labour- and cost-efficient (these are still used in remote regions of the UK with plenty of space). Wrought-iron grids were used to enclose Victorian houses in the UK, and in the Arts and Crafts era palisade fences were fashionable.

A fence may be quicker and cheaper to build than a wall and will produce a more immediate effect than a tall hedge, but it is rarely a long-term solution. Fences lack the rich biodiversity of a hedge, unless you use them as a vertical support for climbing plants such as honeysuckle, roses and ivy. A softwood fence will usually need weatherproofing with paint or a stain, too, so a more sustainable choice would be a long-lasting wood such as oak, cedar or larch that can be left to weather to a beautiful silver colour without treatment. If you do paint a fence, chose a biodegradable product and avoid stains that kill the biodiversity of creatures that live in the wood.

If your fence is only needed for a short while, woven willow or hazel hurdles will last several years before decomposing happily into the soil. These materials are also good choices where budgets are tight and they can prevent unnecessary waste, since fences often risk being ripped out by the next owner.

In the countryside a simple metal estate fence can be left raw to blend in with the landscape and avoid the toxic airborne chemicals used in the powder-coating process. The traditional method of installing these is to thread long bars of metal through individually secured uprights and fix them with clips at each post, allowing the fence to be taken down and reused in the future. In a country setting a wooden post-and-rail fence is also unobtrusive, with its rustic split timbers made from renewable coppiced chestnut. Rails should be morticed into the posts and the posts dug deep into the ground for stability.

Fences also filter wind in a similar way to hedges and can provide good shelter without creating the wind eddies associated with the lee of a wall.

A stock fence can often accommodate a hedge along its stretch as well, and if you want to exclude rabbits, add rabbit netting and inset it into the ground to prevent the creatures digging underneath. However, it is kind to provide a small hole in the bottom of a fence to allow small mammals to forage between gardens, over a large enough territory for them to meet and eat. Badgers can be destructive in a countryside garden, where they will dig under or barrel through most barriers. To protect your fence, install a badger gate along a badger route, which will allow them access, but be too heavy for a rabbit to push open. It may seem strange to impose a hierarchy of animal invitations that welcome badgers and not rabbits, and I know some people who would welcome rabbits into their gardens, and others who would see them as part of the sustainable food chain. The choice is yours.

Horizonal woven chestnut and vertical reclaimed timber make good fences, or why not plant a living 'fedge'?

Materials

Choosing from the many materials available on the market to build a garden, before we even get to plants, can be overwhelming. Decisions need to be made based on budget, durability and looks, as well as sustainability, with some greenwashing to wade through along the way, so it's good to have some guidelines to help you choose materials that are ecologically sound and offer value for money.

Hard landscape materials

You may feel that you are less qualified to choose your materials than your builders, but as these decisions can affect both your own health and the health of the planet it's worth spending time assessing and discussing the options.

Hard landscape materials include stone; aggregates such as gravel, sand and cement; metals; woods; clay products, including bricks, tiles and ceramics; plastics and fabrics. The ideal materials are those that use few resources, have a low environmental impact and present a low health risk to humans and wildlife. They should also help to create a sustainable site and support companies with sustainable practices.

Consider the durability of the material and its fitness for purpose, checking where it will be placed and how it will weather. Clay bricks are very adaptable small building blocks and so minimize waste and resources. They are useful for walls if protected from water penetration, but you will need a harder engineering brick on the ground to prevent 'blowing', which is when products crack after water penetrates the porous clay and expands as it freezes.

By selecting the right materials you can also avoid the waste of repairs and redesign. If you choose a local material, such as bricks in an area of clay soil, they will eventually disintegrate back into the land without trace and so have minimal environmental impact. Wood is a very adaptable material for building, and it is also biodegradable, but if it's stained or painted with a toxic product it can leach contaminants, so choose timber that will last without being treated.

Once you've chosen the design layout for your garden you can specify combinations of materials to help to create a sustainable site. For example, in a typical terrace construction stone can be laid either rigid or loose. Loose means laid on sand over a naturally quarried or recycled aggregate, which allows water to drain and plants to grow in the gaps. The stones can be lifted and recycled and the aggregate below the flagstones can be dug into the soil if the design changes, minimizing waste. Solid construction would include flagstones laid on solid concrete so that

Local is best. Gabions filled with clay blocks from site, a fallen-oak bench and a growing tunnel.

A concrete drain makes a great planter.

they do not move. This impermeable surface may look neat and even, but it will also cause rainwater to run into drains faster, exacerbating the risk of localized flooding, and the cement-secured stones will be difficult to lift without breaking them if the design changes. Plants will not be able to grow between the cracks, reducing biodiversity and contributing more to urban heat islands (when built environments such as cities retain heat), in turn adding to global warming. In addition, the concrete base beneath gives off carbon dioxide when it cures and prevents the soil below from breathing after it's laid. If the design changes, the concrete will need to be broken up and the stones will often break too when it is removed, making them more difficult to recycle. The untrained eye may see little visual difference between these two terraces, but they have widely different effects on the environment.

Sustainable ethical sourcing and reuse

As the construction industry has grown and centralized, building has become less about local materials and craftsmanship and more about mass production, standardization and convenience. The drive to ensure a low price to the end user does not count the real cost of the products we buy and use to build. Highly engineered and complex composites and the liberal use of chemicals to manufacture, preserve and clean materials all have a high cost to the environment. The more complex the material, the more energy is usually needed in its manufacture and the less easy it becomes to disassemble and reuse. The distance a material travels also increases the transport fuel needed, which often accounts for the greatest use of energy in a heavy product's life cycle. Transportation depends on finite resources of fossil fuels, and increases greenhouse gases and pollution, which is why buying local is one of the simplest things we can do to make a difference, especially with heavy construction materials such as stone.

As consumers, we should question where stone is from and how it was extracted from the land. Indian or Chinese sandstone may seem a cheap option, but if you factor in the distance travelled by container ship (with occasionally disastrous oil spills) it may seem a less attractive choice. Also ask suppliers how it was quarried: child labour and harsh conditions have historically been rife. While no equivalent to the FSC (Forest Stewardship Council) certification for wood exists for stone yet, check whether your suppliers belong to the Ethical Trading Initiative (ETI) or Earthworm (formerly the Forest Trust), which inspect and work with some quarries to improve conditions.

The durability of concrete, like stone, makes it an attractive building material, but to avoid the high level of greenhouse gases emitted during its manufacture, we can reduce the proportion of Portland cement used by replacing it with fly ash, for example, a waste product of the coal industry.

Whole life use is important, too, and choosing a product that has been recycled and can be recycled again is the most sustainable option. Even concrete can be reused: hiring a crusher will turn it into aggregate that can be used as a sub-base for a drive or terrace, reducing landfill and transport costs. Plastic is a durable product, although highly reliant on fossil fuels in its manufacture. Composite decking may seem an attractive option because it mixes recycled plastic with wood, but if it is not as feasible to reuse as wood or plastic in their simple state it may end up in landfill.

Aluminium uses a large number of resources to make and may travel between several continents during the various stages of its manufacture before being finished as handrails or light fittings, but it can also be recycled many times at low cost.

A bug hotel made of pallets and wood offcuts provides homes for bees and insects.

Wood

Trees sequester carbon, which is trapped in the timber until it rots down and is released back into the soil. As long as we are building with a wood sourced from trees that grow fast enough to replace themselves in the ecosystem – in other words, faster than the wooden product will decay – it is locking up carbon and reducing greenhouse emissions. The method of procuring wood can be problematic, however, with huge ecological damage caused by clear-felling large tracts of land, leading to soil erosion, loss of biodiversity and wholesale extinction of species, so always choose wood bearing the FSC kite mark, which shows that it was sourced from sustainably managed forests.

The typical life cycle of a product begins with material extracted from the earth and often ends with it being returned via landfill. The aim in a sustainable garden is to change this pattern into a 'closed loop', where products can be reused or recycled many times, and where waste from one process feeds into the production of the next. In addition, we can reduce the quantity of materials we use and the waste we create by making informed choices. If you cannot recognize what a product is made from, it may contain too many different composite ingredients to return gracefully to the soil as it decomposes.

A wood and metal bench beside a table made from a glass windowpane.

Paths – concrete

Boundaries – brick wall

Surfaces – gravel

COMPARING MATERIALS

- £ Unit price per multiplier
- Energy usage
- Resource depletion (organic)
- Resource depletion (inorganic)
- Global warming
- Ozone depletion
- Toxins
- Acid rain
- Photochemical oxidants
- Durability/Ease of maintenance
- Recycling/Reuse/Disposal

It's not easy to decide between materials with so many factors to compare. This is a rough guide to some of the things to weigh up when choosing, with an approximate scoring system I use applied to each. You could build in other factors and refine this table as you go.

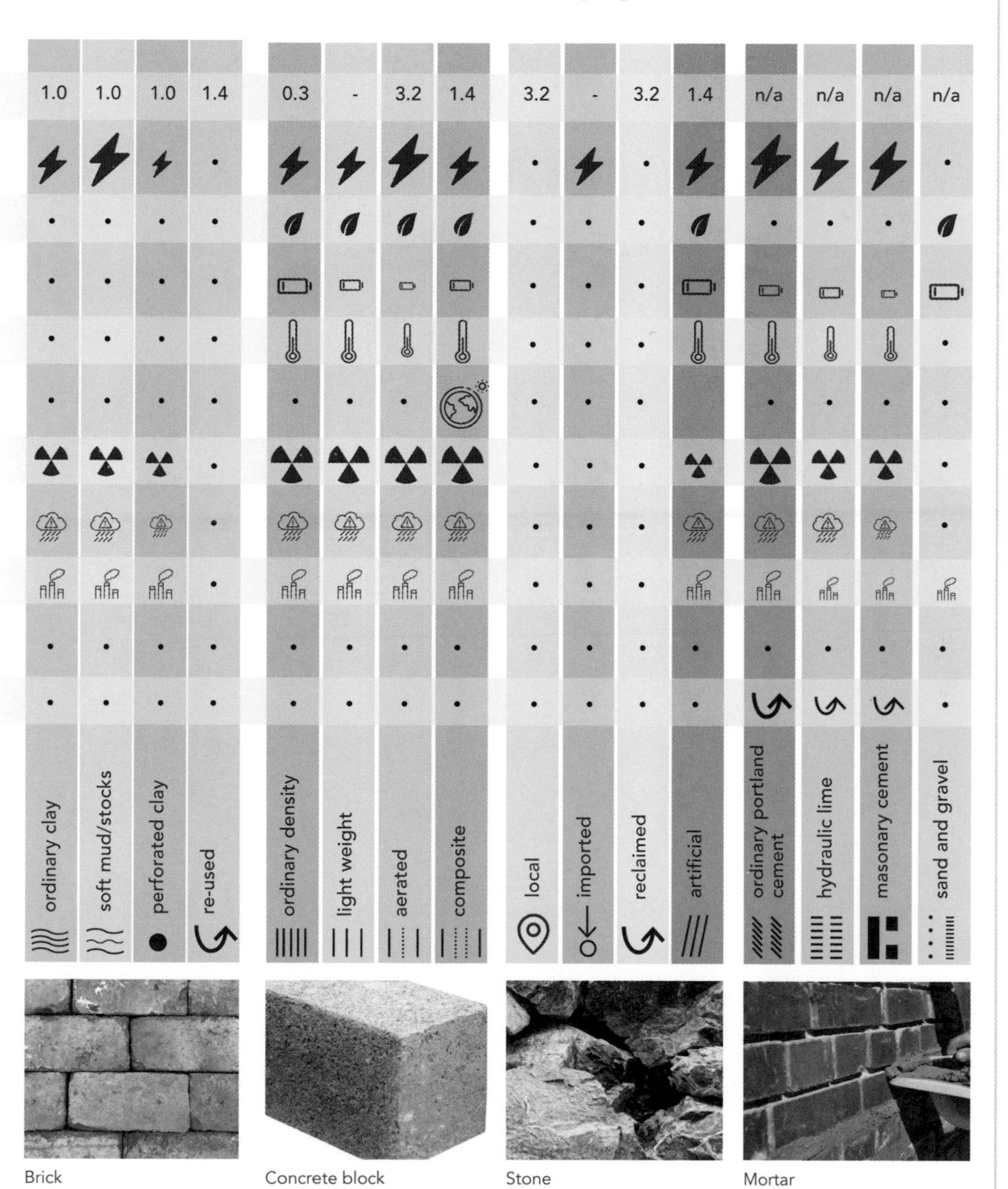

Brick

Concrete block

Stone

Mortar

1
2
3
4

Project: Plant an edible hedge

Planting an edible hedge brings joy into your garden and offers a great way to connect to and notice the seasons for both children and adults. It differs from a normal garden hedge only in the choice of plants and timing of the care. You cut it less often to allow the plants to flower and fruit, something which we can all do anyway to help feed the wild birds as they migrate on their way to and from warmer climes.

Choose a mixture of berries and structural plants and buy bare-root young trees, known as whips, in winter. Hawthorn and blackthorn are the classic base plants that knit the hedge together. Blackthorn has long sharp thorns, so be careful if you have small children. Hazelnuts are delicious raw or will keep for ages when dried. Sea buckthorn is a good choice for cold and coastal sites, its fiery red berries ideal for jams and pies; crab apples can be made into jellies and will feed the birds when all else is eaten in the coldest winter garden; and you can make cordials from elderflowers and a powerful vitamin syrup or wine from the berries.

Materials & plants

- Bare-root whips
- Cardboard
- Compost (home-made ideally)
- Rabbit guards (optional)

Method

To allow berries to develop, the hedge needs a little bit more space than a formal hedge to grow comfortably. Leave 50cm/20in from the edge of the hedge to a boundary or path to allow you access to it for pruning and picking each side.

Apply a layer of thick cardboard, with a 5cm/2in layer of compost mulch on top (*see page 138 for no-dig planting*), over the area where you plan to plant the hedge. Leave through several rains or water it yourself, to soak the cardboard.

When your bare-root whips arrive, soak them in water for a few hours. **[1]** Then move the mulch aside and make holes in the soggy card so that you can plant each whip at a spacing of 50cm/20in in a staggered row. **[2]**

If you have rabbits, protect the whips with rabbit guards or wire netting until the hedge is established. **[3]** Water for the first year during dry spells only.

Harvest leaves in spring and berries in autumn. **[4]** Some berries such as elder are toxic when raw and many need preparation and cooking before eating, so always consult a good hedgerow cookbook before eating.

Project: Plant a dead hedge

In deep shade under trees where the roots are close to the surface and only the smallest holly seedlings can be planted, a dead hedge will protect the plants as they get going. It can also be a legal requirement to lay one on construction sites where live hedgerows have been taken out, until a new hedge is established. These features also provide excellent wildlife corridors between territories for small mammals – you will even find birds of prey perching on them to hunt. As dead hedges decompose, the wood becomes the perfect host for invertebrates and fungi. These hedges also make great barriers and free alternatives to a fence, using cuttings from tree pruning and any woody material that can be laid in rows.

Materials & tools

- String, a hosepipe or small pegs
- Timber posts about 1.5m/5ft long
- Axe
- Mallet or post driver
- Prunings etc.

Method

To begin, mark out the line of the hedge with string or a hosepipe or small pegs. A good location would be between the compost area and the main garden, at the edge of the garden, or in a woodland area where you would like to encourage wildlife.

Drive some vertical posts about 30cm/12in into the ground to create a frame – using an axe sharpen the posts to create pointed ends to make this job easier. Using a mallet, or borrow or hire a post driver if you want to cover a large area and save your shoulders from injury, position the posts in pairs, 90–120cm/3–4ft apart at 1.5m/5ft spacings. [1]

Collect together small branches and twigs, either dead windfall or prunings from trees, live hedge clippings, dried ornamental grasses, or reeds from ponds. If you don't have enough material your neighbours may be happy to give you their garden waste. Pile the brush into the gaps, using the largest pieces at the bottom. [2]

Try to overlap the piles, as though you were laying bricks in a staggered bond, to knit them together. [3]

As the woody material decomposes the hedge will sink, so you can add to it each year, or grow ivy over it and allow it to gently break down.

1

2

3

Starting with soil

For plants to grow, they need access to light, air, water and nutrients. The ground beneath our feet provides the anchor to hold them steady, allowing their stems to reach up to the light and air, while the soil around their roots also gives them access to water and nutrients. But soil is so much more than a useful growing medium. The number of living organisms in just a teaspoon of it is greater than the number of humans in the world. Each square centimetre of soil is teeming with microorganisms, bacteria and fungi that work together to break down dead organic matter and create new life; it's as simple and as complex as that.

Soil is our best defence against long-term climate change as well, and has a greater ability to lock up carbon than even trees. We are the soil we eat: the soil that our food grows in and the soil that our bodies will become when we have finished with them. If that's sobering, it is also worth thinking about the extraordinary properties of soil in an extreme situation, such as the aftermath of the Chernobyl disaster. The nuclear explosion poisoned everything around it, and humans could not live there safely, but after very few years the soil had begun to heal the land and the first fungi, then plants could grow again.

The best way to create a sustainable soil is to protect it. Avoid poisons, avoid pollutants, and avoid squashing the life out of the soil with heavy machinery. In the past, our over-use of pesticides and fertilizers has left the soil at best depleted and hungry and at worst poisoned, as farmers sought higher yields and gardeners aimed for brighter blooms, greener grass and fewer 'weeds' by adding cocktails of chemicals to the ground. We are now understanding that nature is very good at regulating itself, and our job is not to control, but to assist in the natural process of regeneration by maintaining soil health. The way we plan our gardens can be as helpful as the way we garden them, from protecting the soil to building in a circular soil economy and keeping the benefits on-site.

Healthy soil smells deliciously earthy and wholesome! Putting our hands in it releases endorphins which make us feel good too.

Identifying your soil

As well as organic matter, your soil will contain different minerals to make up its structure. This structure, built up over many years, may have more chalk, clay, loam, sand or silt, depending on the base rock below. Each of these structures will hold varying amounts of water and nutrients and be more or less acidic, and each will favour the plants that naturally grow in sites with these conditions. To get to know your soil and choose plants that will thrive in it, take a handful and add some water, then roll it in your palm. If it forms a firm sausage or you could make a cup out of it, you have clay; if it is gritty, it will be sandy; and if it is silky, it is likely to be silt. You can also buy kits to easily test the pH level of your soil. If you have a large garden. test in a few places and choose the right plants to suit the local conditions.

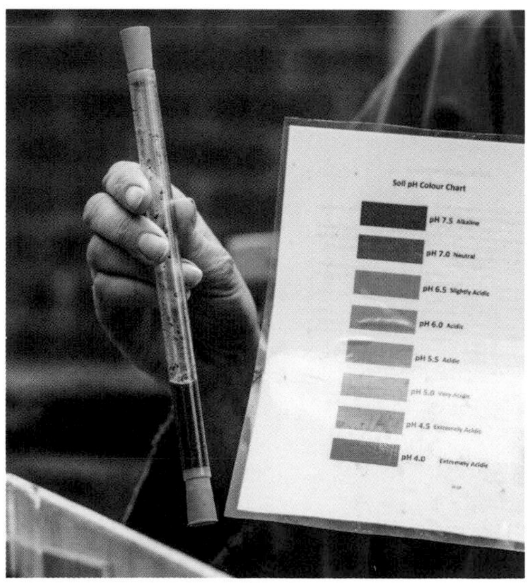

A simple soil-testing kit will tell you the pH.

Soil health

By planting in layers with ground-hugging plants at the bottom, followed by small shrubs and perennials, and finally shrubs and trees for height and structure, and covering any exposed soil between them with an organic mulch, we help to keep carbon locked up and prevent nutrients being washed away by heavy rain or flooding. The mulch will gradually be taken down into the soil by worms and the nutrients it contains broken down to feed the plants, so it's a good idea to choose a mulch to suit your soil.

If you garden on heavy clay, a mushroom compost is a good choice, as it will help to break down the clay by opening up channels between the densely packed particles. If you grow roses, a mushroom compost will be too alkaline – they prefer a rich, well-rotted manure, which has a higher nitrogen content and is a little more acidic but will still help clay to drain more efficiently. If you garden on sand, a well-rotted manure will add nutrients and help the free-draining particles to retain more moisture. If you don't want to delve too much into the pH value (acidity or alkalinity) of what you are adding to your garden, you can't go far wrong with a well-rotted garden compost, either your own, or purchased from locally recycled green waste. The great benefit of creating your own is that you know what has gone into it and can avoid including glyphosate weedkillers and chemical pesticides. Many composts that you can buy have no such guarantee and it is therefore perhaps best to use them on the ornamental garden and not the vegetable plot, where whatever is in the soil will be returned to your plate.

A healthy soil will be full of roots and a network of mycorrhizal fungi that work together to distribute sugars and nutrients to plants and even convey messages about attacks from insects or diseases. We should protect those networks by digging as little as possible (*see page 138 for more about no-dig*) and feeding the soil from the top down.

The price of peat

One way of improving our gardens' sustainability is to avoid the use of peat. Although steps are being taken to reduce or eliminate it from gardening products, in the UK 2.1 million cubic metres/2.7 million cubic yards each year was being used in the domestic garden market up until 2019.

The peat used is stripped from rare bogs and transported mostly from Ireland and Eastern Europe and the cost to the environment is huge, from the habitat destruction and biodiversity we are removing from source, to the thousands of years it will take to replace. Peat bogs are important carbon sinks, too, so every pot plant full of peat contributes materially to climate change.

So why do we use it? In the 1960s it was found to be a stable and steady material in which to grow plants; it also holds water well and is not too heavy to transport. It makes a good soil conditioner, opening up the structure of a clay soil and adding nutrients to a sandy soil. Early formulas for peat-free growing media were less stable and got a bad press for growing less impressive plants, putting people off. Now that the science is better understood, it is much easier to find bags of good quality, peat-free compost at the garden store to grow seeds and take cuttings, and for everything else, home-made compost is the best option.

Feeding the soil

Soil feeds plants with minerals and plants return their nutrients to the soil when they die and are broken down by creatures such as beetles, worms, fungi and bacteria, after which they are taken back up again by new plants. It's an incredibly efficient cycle and, as gardeners, the worst thing we can do is to get in the way by being too tidy. Pruning, weeding and clearing away deprives the soil of the dead twigs, leaves and branches it would usually accumulate to break down in the normal cycle in the wild. To keep our soil full of nutrients we have to replace the biomass we take away when we tidy up.

Returning goodness to the soil through compost.

We can send off our plant waste to the local recycling centre in a roadside bin and buy it back as compost, but it makes sense to transport it as little as possible to minimize fossil fuel usage, so if you have space, keep it on-site.

Your garden waste will soon break down in a compost heap, wormery or in a pile of leaves. Most natural materials we throw away will break down on the compost heap, from prunings and leaves to tea bags and vegetable peelings from the kitchen, to paper bags and newspapers. You can even add cotton, wool or viscose clothes that are beyond repair. Entire books on making compost are available, but the key is to balance your browns (carbohydrate-rich materials) and greens (nitrogen-rich materials), and to turn the contents every now and then (if you have the stamina) to hasten the process.

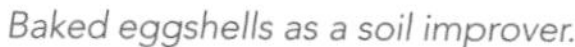
Baked eggshells as a soil improver.

Make your own seed blocks.

Help the soil cycle

Some human food attracts animals such as rats and foxes, so use a wormery for cooked waste if you want to avoid night-time visitors (*see page 67*). Eggshells are alkaline and make a great addition to a clay soil, but they take a long time to break down in the wormery or compost heap, so to stop them attracting animals, I bake them first, then crush them and sprinkle them over the soil as an improver and slug repellent.

Coffee granules are slightly acidic and can be thrown straight on to the borders around roses and other plants that like these conditions, where they too act as a soil improver and natural slug repellent. You can also add them to water, diluting them in a ratio of one-part coffee to ten-parts water, to make an occasional nitrogen feed for house plants.

Another great way to feed the soil is with green manures. For the price of a few seeds, you can sow a crop of plants that will increase fertility and improve soil structure, either on bare soil or between smaller plantings. Plants such as clover and legumes (from the pea and bean family) fix the nitrogen that is freely available in the air in their root nodules, releasing it later for nitrogen-hungry plants to take up. Other plants like buckwheat have deep roots that seek out nutrients and bring them closer to the surface for new plantings to take up.

Green manures such as phacelia can also be great bee forage, providing nectar above ground and nitrogen below. Simply cut this beautiful plant down before it sets seed to stop it spreading and either dig in the flowers or leave them on the surface to decompose.

Coffee granules can be sprinkled straight on to the soil.

1

2

3

4

Project: Make your own comfrey tea

Comfrey tea is deeply satisfying to make and couldn't be easier. With little effort you can make a nutritious feed for plants containing all the nitrogen of the comfrey leaf. Comfrey tea is perfect for tomatoes or any other plant needing a bit of a nitrogen boost, which is the case when growing fast and making new leaves. Nettles can be used in the same way, and curiously, biodynamic gardeners also use the tomato leaves, cutting the lower ones from the plants to make the tea. I have followed this practice and avoided the ubiquitous tomato blight, so while I don't yet understand the science, I am a fan.

Materials

- Bucket or bin
- Comfrey leaves (*Symphytum officianale*)
- Old compost bag or lid

Method

Add water to a bucket or old rubbish bin so that it is two-thirds full.

Pick and then immerse comfrey leaves and flowers, topping up the bucket with them to arrive at a ratio of two-parts water to one-part leaves. **[1–2]**

Cover the bucket – a lid or an old compost bag is ideal for this. **[3]**

Speed things up by leaving the bucket in a warm place for about two weeks.

Once it smells so terrible that you cannot quite believe it will be good for the plants, dilute it to one-part comfrey tea to five-parts water, and water into the soil around your plants, avoiding the leaves in case they scorch in sunlight. **[4]**

Project: Make a leaf compost bin

There is little benefit in using a machine to tidy up leaves from a garden, pack them in bags and leave them to be taken away, along with all those free nutrients and all that free soil conditioner.

Leaves are full of the sugars made by the tree during the year and then dropped in autumn when they can no longer photosynthesize. They break down quickly and easily and within one year can provide a lovely leafmould to enrich your soil. In a wooded area you can just leave them where they have fallen; it's sweet to see the snowdrops and other spring plants grow up through them. But if space is at a premium, or just to tidy up leaves on a terrace or lawn, a compost bin is the answer.

.

The aim is to hold the leaves in one place while they rot down, and old compost bags or reused builders' bags with holes pierced into the sides do the job perfectly. A squared off area made from chicken wire or old pallets attached to wooden posts will also work, while hop or apple bins make great ready-made containers.

If you don't have a handy skip or farm nearby to salvage a bin you will need four wooden posts and a roll of chicken wire to create a compost bay.

Materials & tools

- Timber posts about 1.5m/5ft long
- Axe to sharpen ends
- Mallet or post driver
- Roll of chicken wire
- Staples to fix to posts
- Hammer
- Pliers

Method

Drive some vertical posts about 30cm/12in into the ground to create a frame – using an axe, sharpen the posts to form pointed ends to make this job easier.

Using a mallet, or borrow or hire a post driver to save your shoulders from injury, position the posts in a square, 90–120cm/3–4ft apart. **[1]**

Attach chicken wire to each post using steel staples. **[2]**

Cut the wire to be long enough to reach around the whole square so you can close the fourth side once it is full. **[3]**

Fill the space with leaves and keep topping up as they fall, then close the front and leave them to rot down. **[4]**

1

2

3

4

The wonder of worms

The best example of the closed loop economy, which is based on the principles of designing out waste and pollution, is to remove our waste by feeding it to other creatures, and then to use their waste to feed ourselves. In my house there is a hierarchy of leftovers, which if they can't be reused, are sorted to be given to the dog, the chickens or the worms, depending on what they contain. Cooked food and meats are added to the dog food, while bread or vegetable scraps, if too old for vegetable stock (or after they have been used as stock), go to the chickens. Cooked food and bones, tea bags and vegetable skins or husks (that are no good for dogs or chickens) go into the wormery. Leave out lemon and orange peel, as they are too acidic and will turn the wormery mixture sour and attract flies. (*See page 60 for uses for coffee granules and eggshells.*)

The wormery has the benefit of being rat- and mouse-proof because it is covered and has legs, making it a good choice for a smaller garden. It can sit right outside the back door as it does not smell either, when well balanced.

A wormery is also easy to set up. It's based on a tiered system of trays with holes in the bottom of each one. These holes allow the worms in the tier above to eat the newest composting food and their waste to fall through and collect in the lower tiers. The waste is full of nutrients and with a tap system the thick dark liquid can be collected and added to the watering can. When the bottom tray is full of rich dark waste it can be emptied on to the vegetable garden or into your prize pots.

A wormery is a virtuous loop, turning food waste into nutrients to create more food.

Water, water

Water is a finite resource. The water we have has been with us since the world began. Every glass of water we drink is likely to have been drunk by people before us, and possibly a few dinosaurs. The earth does an amazing job of recycling water through rain and filtration in deep aquifers under the ground, but by overusing water we have put a burden on the system. Clearing land of trees has created barren deserts and by treating our rivers and seas like sewers we are poisoning our very life source.

Seventy per cent of the earth's surface is covered in water, and our bodies are 70 per cent water, and 90 per cent of all animal life is found in water. So it's not surprising that water has long been held as sacred, yet in most first-world countries where we have been lucky enough to have plenty of water historically, we have taken water for granted in recent times.

A properly designed ecosystem builds water in from the start. Saving water is not just about emptying your bathtub on a bed of begonias. We have to be careful with every drop of water that comes into our home and garden from the very start. The one thing we cannot design in the garden, though, is how much rain will fall. Even in the temperate regions we have had long periods of drought followed by the worst flooding in living memory. As the weather becomes more extreme we can expect these fluctuations, and we have to plan for them. But we can also help to mitigate them by what we do in our homes and gardens.

Preserving water

When it rains: if you live in a city the more you can do to slow the flow of water the better. Water that falls on the roof can be caught in butts and ponds, catch pits and underground tanks, both to use later and also to prevent the community drainage being overwhelmed causing flooding and sewage to overflow into the rivers.

The best place to store water is of course in the ground. By having permeable terraces and paving and allowing water to run into the earth it can by filtered by natural aquifers. Trees and plants then take up the water and return it to our ecosystem. Forests have even been shown to increase the chance of rain when rain is needed so of course the best thing you can do for the long-term health of our water is to plant trees.

When it doesn't rain: cover the earth with planting and mulch to protect the soil and help prevent evaporation as well as carbon dioxide leaching into the atmosphere. Choose plants that suit your soil so that they don't need too much water. Plant in layers to create a mini ecosystem, and introduce a water feature into your garden to help conserve water, as a pond creates its own mini microclimate. In the next section we will look at reusing rainwater and making our gardens into a living swale.

Collecting and reusing water

Every drop of water we reuse in the garden prevents it from going down a long drain to be treated and returned to us via another long pipe, using more energy. A water butt is easy to attach to a downpipe from the house or glasshouse and can overflow on to the garden or into drains during periods of heavy rain. Place it on bricks high enough to fit a watering can under the tap,

Collect rainwater in ponds or butts to use on the garden.

Washing-up water using eco-detergent can be used too.

or set it at a level where you can take the lid off and dip the can in from above. In larger gardens you can install rainwater harvesting tanks buried below ground, with pumps, automatic top-ups and timers to feed irrigation systems. If you need to irrigate your garden, use a leaky hose, which will waste less water than a sprinkler that loses water into the air, and install tree bags, which can be fitted around trees and filled with water manually at regular intervals to seep through gradually and ensure only the correct amount of water is used. These measures will all protect water supplies, but if you design your garden carefully you will only need to irrigate plants while they are establishing or in extreme drought.

Rainwater will collect bird droppings and other debris on the way from the roof to the water butt or tank, so it is not suitable to drink untreated, but plants and soil will be delighted with these additions. The water can also be used for flushing lavatories and washing clothes, but you'll need to install a filtration system first.

To minimize your water-poisoning footprint reduce the chemicals and plastic you flush down the lavatory and via the washing machine. Plastic fibres have been found in every ocean and in our fish and sea creatures, so buy clothes that are made of natural materials. The simplest way to recycle grey water is to wash up in a bowl using an eco-friendly detergent, and then tip it on to a mulched border to filter through to the garden. This can be taken one stage further by diverting grey water via a filter bed. This bed is layered with stones and soil and planted with wetland plants such as bulrushes (*Typha*) and canna lilies that will cope with and filter the detergent. It is important to add a layer of gravel as a top dressing and keep the water below this level if you want to avoid mosquitoes laying eggs in standing water. Although beyond the scope of this book, you will find designs for filter beds on the internet.

Water for wildlife

Even in the smallest of gardens a bird bath or shallow pot of water will draw in birds and insects, and if you create very shallow areas you will see bees drinking from them. If you have more space, vintage tubs or baths set around the garden can collect water close to thirsty plants, making watering easier and allowing visiting mammals to have a drink too. Don't forget to add shallow areas and slopes at the edges using bricks or pieces of wood to allow smaller mammals and invertebrates to get in and out.

If you have a large enough garden a pond is fairly easy to install and will take your biodiversity and wildlife habitat to another level. It will also change the microclimate in the garden, providing moisture as the water evaporates, thereby cooling the air on a summer's day and keeping it a little warmer in the winter.

A source of water in a garden benefits all wildlife.

If your garden lies on clay, you may be able to create a pond lining by puddling the soil. This involves smoothing the surface of the clay to create an impermeable layer, like making a cup or bowl. Farmers historically used the large hooves of cattle followed by the smaller hooves of sheep to achieve this, so you might tread it by foot or use a roller, or, on a large scale, a mechanical digger with wheels.

If your soil is chalk or sand, you will need to line the pond. Synthetic rubber liners such as EPDM are non-toxic and do not break down in sunlight, and they will have a lifetime guarantee. These liners used to rely on fossil fuels in their manufacture, but more recently sugar-cane-derived ethanol has been used to create the ethylene. Alternatively, use a 10in/25cm layer of powdered bentonite clay, a natural product, on the base and sides of the pond and ram it down, then apply a layer of soil mixed with the clay, which you also compact firmly. Remove stones and topsoil from the bottom of the pit before adding the clay, and after the second layer of clay, lay coconut fibre on top to protect the soil edges.

After lining your pond, use stones and rocks around the edges to create habitats for invertebrates, and finally plant with oxygenators and marginals in baskets or in soil wrapped in hessian to create a healthy ecosystem and keep the water clear. Aim to shade 50 per cent of the surface with plants to avoid algal bloom, which can appear in warm water that lacks oxygen.

A simpler way to provide water for wildlife is with a swale or ditch that will fill with water when it rains and then drain away. Simply follow the natural contours of the land by watching where rain flows downhill and dig a shallow trench in the path of the water. Then plant the trench and the areas around the sides with a wildflower and grasses mix suitable for your soil.

Project: Make a simple rain garden

To make a rain garden, you will need a solid or semi-permeable surface so that water will collect there when it rains. An old trough or stone sink is ideal, but even a recycled compost bag will do for a small area. You can cover the base with stones and pebbles to create different levels and then plant up with small water-loving plants or leave it empty to be used by birds and insects to drink.

Materials

- Old trough or stone sink
- Stones or pebbles to cover the base

Method

The rain garden should fill when it rains and just drain or overflow during a heavy downpour, so place it on the soil or a permeable surface to allow the water to return to the earth. The idea is to slow the water down and catch it, so that you can enjoy it for a while without creating a permanent pond.

These little gardens can turn a rainy day into an event, so put one near a window if you want to watch from the comfort of a dry room. The sound of the rain will change as it hits the different levels of water and the edges of the stone.

Plastic: from love to hate

Plastics are in our soils, our oceans, our food and our blood. The giant accumulation of plastic in the ocean dubbed the Great Pacific Garbage Patch contains at least 71,000 tonnes/79,000 tons of discarded plastic and covers an area of about 1.6 million square kilometres/617,800 square miles, while tiny particles or microplastics have been found across the globe in many food products, including 90 per cent of table salt brands. We have suddenly woken up to the nightmare of poisoning ourselves through our own cleverness. Because plastic is really clever. A malleable, shape-shifting medium made primarily of oil that can bend and stretch and be thin or strong or both. No wonder we fell for it when the first single-use plastic bag was produced in 1965.

But after the love affair, the backlash has been almost as bad, with people discarding all things plastic to own something more 'sustainable'. Throwing stuff into landfill to buy something else isn't going to help, though. If we have plastic, whether plant pots, tables and chairs or polytunnels, we should carry on using it. And repair it and use again, until it totally falls apart, by which time technology may have caught up, allowing us to recycle more plastics.

It's a shame that garden compost is usually sold in plastic bags. If you are buying in quantity, ask for your garden supplies to be delivered loose in bulk, or in 1 tonne/ ton reusable sacks, but if you have bought compost bags keep them to reuse. They are usually black inside which makes them ideal for creating a mini rain garden or popping over a bucket of comfrey tea, turned the black side out to absorb the sun's warmth.

Plastic bottles make good bird feeders when tied upside down, or cut them in half to create perfect mini cloches to protect seedlings from the birds and early frosts. Alternatively, cut the bottom off a bottle to create a funnel to water tomatoes and other glasshouse plants; half bury one in each pot to allow the water to reach deeper and slow down the flow to reduce run-off.

As the old joke about asking for directions goes, 'I wouldn't have started from here', but now we have plastics let's use them and from now on only replace them with alternatives when they perish.

A new play area made of reclaimed stone and planting; plastic can be reused as mini greenhouses, or to patch a bigger one.

Plastic alternatives

The good news is that the plastic issue is being tackled and alternatives are becoming available. As consumers we can speed up the change by asking questions and choosing sustainable products instead of plastic pots, tools and packaging wherever possible.

The most visible plastics in the garden are plant pots. Lightweight and non-porous, plastic would be a tough act to follow if it wasn't the silent killer of the seas. Black plastic pots cannot be recycled but taupe plastic pots can. They are made from carbon black-free recycled polypropylene and are identifiable by near infrared (NIR) as recyclable by the machines that sift domestic waste. Better still, terracotta pots are made of natural clay quarried from the ground and will last for many years. Beautiful and naturally porous, they hold a little water and keep plants cooler than other materials in summer, while offering some protection to roots in winter. To prevent your clay pots and plants drying out, use capillary matting in the glasshouse and stand them on saucers to water from below in the house.

Once plants are in the ground, stored pots usually add to the messy corner of the garden, unless they are biodegradable. Biodegradable pots are planted with the plant, so they are not recycled but reincarnated to become part of the soil and so feed the plants as they grow.

Biodegradable options are on the increase and currently include coir pots made from the husks of coconuts; Vipots made from rice or grain hulls; and containers created with cardboard or even coffee grounds. Of these, coir will have the environmental cost of transportation by ship to offset, and coffee is also imported, but once drunk the grounds are here as a waste product so it's great to reuse them. Coir holds some water, so allow for that when watering and keep it moist, since it takes a while to re-wet when it dries out.

Wooden tools feel warm to the touch and wear down gracefully over time, eventually returning to the soil to release the carbon they lock up during their lifetime (*see page 48*). When plastic handles break beyond repair, replace them with wooden ones to give the tools a second life, and choose metal watering cans and galvanized or wooden water butts.

Finally, insist on plastic-free packaging by voting with your wallet. Choose compost that's delivered loose in bulk or in refillable dumpy bags, or in cardboard boxes with a reusable liner for small amounts. Buy plants bare-root, or wrapped in newspaper, or better still make your own compost and grow plants for free (*see page 122*).

Plastic-free pots look fabulous in all shapes and materials.

Recycling materials

The best way to recycle, if you have the space, is at home, avoiding transportation and the energy needed to process material on an industrial scale. Newspapers, envelopes, wooden boxes, cardboard, eggshells, coffee grounds, glass, plastics and organic fabrics can all be reused in the garden.

In the glasshouse:
- Sow seeds in wooden trays or recycled cardboard food cartons.
- Use lavatory rolls for sowing peas and beans which dislike root disturbance.
- Make your own seed pots from newspaper using a potter.

Cardboard tubes make great pots because while standing in water the sides last longer than the bottoms, providing natural air pruning and allowing the roots to get down once they are planted out. You can make your own with cardboard and paper tape or, if you're feeling creative, make a pot out of recycled coffee granules and flour.

Horticultural fleece is lightweight and reusable until it perishes, but because it is made of spun polypropylene it shreds easily and will end up contributing to the microplastics in the soil and water. Instead, use hessian, newspaper, cardboard or straw to protect plants from frost, or try wool, which is a great insulator, retaining heat and moisture before gradually biodegrading.

On a larger scale, a glass cold frame will last longer than a plastic one and, if you have the space, a few bales of straw and an old windowpane make a perfect seedling hotel.

Food cartons and paper towel tubes make great seed pots.

A cold frame made of waste plastic and wood.

Project: Make your own besom broom

You don't have to be a Harry Potter fan to enjoy the simple grace of a besom broom. Easy to make and a joy to use, it doubles as a terrace cleaner and lawn sweeper, clearing leaves from grass with a motion to keep you fit as well as tidy.

To make the broom you will need some twiggy sticks such as hazel or birch and some long winter prunings of wisteria to use as ties, although garden twine or wire will also work. Hazel and chestnut poles, which grow long and straight after about their third year, can be used to make the broom handle, or you can recycle an old mop handle if you have one.

A bunch of the whippy stems from a hazel or birch tree make a good brush to catch leaves and debris; ensure they are pliable enough to bend a little to catch a pile of leaves, but stiff enough to sweep them along to collect. Make sure you store the broom upside down to prevent the ends bending and keep it in a cool shed to prevent them becoming brittle.

Materials

- Twiggy sticks about 90cm/3ft long
- Wooden pole
- Wisteria prunings, garden twine or wire

Method

Cut a straight pole to suit your height, ideally to come up to about your chest.

Cut the whippy branchlets of hazel or birch in winter, before the catkins are out, and wisteria shoots in midwinter for the long and whippy stems needed to make a strong tie.

Gather the branchlets together in a bunch large enough to make the brush, with the cut ends at the top. [1]

Use the wisteria shoots, twine or wire, as a rope to secure the branchlets to the pole. [2] Wrap the binding around the pole six to eight times and then tuck the end into itself to hold it tight. [3] Trim the longer pieces of the brush to make the head the size you want.

Store it upside down and you're ready to fly! [4]

1

2

3

4

Recycled York stone table, eclectic benches and a welcoming jug of flowers make an extra-friendly setting.

Gathering

A garden can be welcoming and warm and have an energy that invites people in, or it can feel cool and aloof, a place to glance at or hurry through and leave. So, what makes a garden the hearth of the house? The simple answer is the people and, more specifically, their intention, which can be felt through every design layer.

When I think of my favourite gardens, they have a real sense of the people that live there and their evolution, because gardens evolve with people. Whether you are starting out with a first home of your own or thinking about how to make an old garden more welcoming, as the seasons pass you can enjoy the process of making that time into memories to share.

If your intention is to welcome people, that will shine through every detail, and your garden will become alive with friends and animals who may not want to leave.

A sustainable garden does not have to be years old to have this timeless quality. By choosing thoughtfully what to put in it, you can gather in the energy of that comfortable lived-in feeling, making family welcome and encouraging friends to linger, while treading lightly on the Earth at the same time. How much more alive and inviting a garden feels when filled with plants and an eclectic selection of garden furniture and structures, rather than a terrace of bland plastic mass-produced seating.

In this chapter, we will look at how to choose materials that don't cost the earth and add to the gentle art of welcoming, making both your garden and your visitors feel as if they have always belonged.

Ageing is a beautiful process to be enjoyed in all its stages.

Making memories

The time we take to put our gardens together is nothing compared to the time it takes for the Earth to make all the raw materials needed to create them. From the millennia to make a piece of stone or plastic to the hundreds of years to make a mature oak tree, when building future memories I like to think of the material memory of each item in the garden to make sure I value it appropriately. I also check that the way each feature is constructed will do as little harm as possible, that it will age gracefully and can be reused or recycled afterwards.

In choosing what to use or buy I also like to consider whether the material carries an energy that I want to surround myself with. From a social sustainability perspective, if your Indian sandstone was the cheapest choice for you in money terms but cost the life of a child in a faraway labour camp, that may not be such a pleasant energy to bring into your garden. Child labour is endemic in Indian quarries – UNICEF estimates that nearly 20 per cent of the workers in a typical Indian quarry are children, some bonded to gangmasters and working to pay off their family debt. Some companies are ensuring that their supply chains are part of the Ethical Trading Initiative, so we can help by checking that any new products come under this scheme.

Unfortunately, relying on products with the words 'eco' or 'green' as a prefix is not enough to avoid 'greenwashing' or false claims of sustainability. When buying new, we need to spend some time understanding exactly what we are purchasing and consider what each part of our potential garden is made from. Many pieces of garden furniture are comprised of several components and we need to be sure about all of their credentials.

Finally, when making memories, why not start with a few? Old pre-loved materials and furniture add instant character to a garden with their patina and wear over time. The Japanese embrace the notion of 'wabi-sabi' or ageing gracefully in design, and we can take a leaf out of their book to improve our gardens.

If buying new, invest in furniture that will last and can be mended.

Choosing furniture

If you are buying new furniture, take time to choose pieces that will last, that will age elegantly and that are made to be mended, so that they can be passed on to another generation of owners and finally recycled. Let's take buying a swing seat with a stainless-steel frame as an example. Steel is made of iron mixed with carbon and recycled metals (to put it simply) and requires a huge amount of energy to produce, mostly powered by fossil fuels, releasing greenhouse gases and generating waste. On the plus side, it will not rust, will last a long time and can be recycled at the end of its life. Sustainable manufacturers will have also made it easy to disassemble to optimize future repair and reuse.

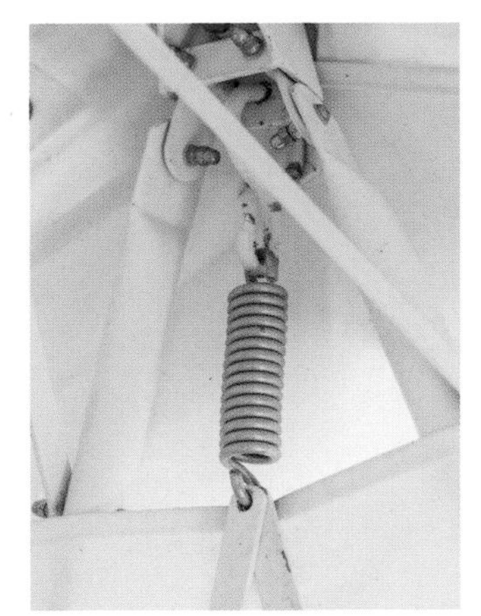

Steel

We can ensure that steel products we buy have:

- Bolted connections rather than welded joints to allow the structure to be dismantled during deconstruction.
- Standard connection details, including bolt sizes and the spacing of holes.
- Easy and permanent access to connections.
- Steel that is free from coatings or coverings that will prevent visual assessment of the condition of the steel or hamper recycling.
- A minimal number of fixings to structural steel elements that require welding, drilling holes, or fixing with Hilti nails; using clamped fittings where possible.
- Components with clearly identified origins and properties (for example, by barcoding, e-tagging or stamping), allowing us to keep an inventory of products for spares.

Once we have assessed the frame, we need to take a look at the upholstery materials. Check whether the cushions are pocket sprung, rather than filled with foam, which is made of polyurethane from the petroleum industry and generally contains toxic VOCs (volatile organic compounds). These are solvents that are released into the air and have been related to health issues ranging from asthma to cancer. Polyurethane is often further coated in flame-retardant chemicals, which have been linked to many illnesses. Sustainable alternatives are organic wool, coconut fibre foam, natural latex or horsehair, the latter sometimes rubberized (choose natural rather than silicone rubber) to ensure longevity and enable reuse.

Finally, check what the covers are made from: they may be cotton sailcloth if you are lucky or, more frequently, polyester. On the upside polyester lasts a long time and can be left outside and washed. On the downside it can only currently be downcycled, in other words, turned into a lower grade material to make items such as swimsuits, which can then become sofa stuffing but could finally end up being burned or in landfill.

By reusing polyester we are delaying the landfill and microplastic problem but not solving it. Hopefully, this will change as technology advances, but for now the best solution is to ask questions and make informed choices, and to take good care of any polyester items you have to make them last.

Check components can be fixed or replaced if they break.

A solid metal framework built to last, with simple wood we can replace.

Buying furniture made locally by individual craftspeople or on a small manufacturing scale allows you to ask these questions directly. They also usually involve less energy consumption in the process and have a lower transport impact than products mass-produced in large factories in far-off places.

Upcycling furniture

If furniture has been made to endure it may even outlast its owner and find its way to a reclamation yard, antiques fair, boot fair or garage sale. These are treasure troves of beautiful pieces to set the scene for a garden retreat or a party place.

There are several antique fairs each year where local sellers or groups of international dealers gather to redistribute the pieces they have acquired, and these events are also now held online via Salvoweb in the UK. Alternatively, check eBay and freecycle websites, and don't forget to also look in the local papers and in skips.

The main thing to look for when hunting for furniture is a solid framework, as so much else can be mended or added to. The timber slats of a park bench may have rotted, but the legs and arms could be cast iron, which will outlast a human life, making it well worth replacing the wood. Old sewing machine bases also make great tables and if the top has rotted you could replace it with anything from a door to a piece of glass, slate or marble – whatever you can find. Once you get your eye in, the combinations are endless.

Watch out for rusty metal nails or snags sticking out of wooden furniture and bang them in or take them out. You are unlikely to come across

With a coat of paint and a pile of cushions, this French day bed will be the perfect seat.

lead paint but if you do, it should be avoided as it's toxic to humans and the soil. Wicker furniture is best kept inside as it decays quite quickly when left out in wet weather. Also beware of faux recycled furniture. You may come across some dainty looking bistro tables and chairs at antique fairs or markets, which on close inspection turn out to be mass-produced and shipped from a long way away. If in doubt, look at the fixings for clues as to an article's age and provenance (*see page 87*).

Reclaiming style

The key to the eclectic look is layering and playing with a diversity of textures and forms. To create a harmonious and calm design, you could keep the same genre of furniture, such as French vintage: an old day bed will look perfectly at home near a repurposed archway from another garden, with a rusted lantern from a local antiques fair making a stunning centrepiece on the bistro table.

You can also give a set of furniture a new lease of life by painting it. Choose one gentle tone for everything to create a restful yin energy vibe, or mix it up with vibrant colours for some extra yang.

When buying paint choose one with minimal VOCs (*see page 87*), which are released into the air as the paint dries. Check the label to ensure that your brushes can be washed in water rather than white spirit and that the paint itself is porous and contains no plastic. Also make sure that your paint is free from formaldehyde and heavy metals, and that it is cruelty-free. The most sustainable paints are made of products that have a plant or mineral origin such as linseed oil, chalk, earth and mineral pigments, and a highly opaque paint will also need fewer coats. Some are even manufactured using hydro, wind and geothermal

Layered brocante and bursting borders make a lovely spot for tea.

power, so shop around for a colour and product with a low environmental impact that will bring new life to old furniture and endure.

Scented plants, candles, textiles and cushions will elevate any setting and the joy of reclaimed textiles is not minding too much if they get weathered. A Khadi fabric is made of many pieces of material oversewn to create a quilt effect, each one a unique amalgam of memories. You can make your own simply by layering old sheets, shirts and dresses and sewing them together in lines or patterns. For a more formal look take a woollen rug and add an old chintz cushion to bring the drawing room outside.

Reclaiming function

Great garden furniture does not have to start out as furniture. Try repurposing other items as well to create the look and set the scene for your own personal haven. Wooden wine boxes are perfect for growing herbs, storing logs or housing your gardening kit. A galvanized metal bathtub or bucket makes a perfect mini pond with just a few plants, or drill some holes in the bottom to create a chic city planter. Food and eco-paint tins can also become plant pots, while a ladder leant against a wall can double as a plant support – perfect for a rented home where holes can't be drilled into walls. Some materials such as concrete come with a high carbon cost, so when items like large drainpipes are surplus to requirements, it makes sense to reuse them.

To make extra seats, wrap bales of straw in hessian or burlap, which is woven from natural fibres derived from the jute plant. Hessian is breathable and does not rot the straw with condensation; just keep the seats under cover in heavy rain or allow them to dry out in the sun before use. Draped in rugs and cushions these seats make a super, convivial and flexible gathering space around a campfire. Be careful of stray sparks, of course, as natural fabric is flammable. At the end of the season the straw can be added to the compost heap or used as mulch in the vegetable garden, while the hessian can be kept for the following year or used to protect tender plants from hard frosts over winter. I also use it to wrap larger recycled plastic pots to bring into the studio.

Furniture can be made of an old bathtub, a pallet or a bale of straw.

Project: Use recycled planters

Old, galvanized bathtubs make great planters. Before indoor plumbing was invented bathtubs, like chamber pots and washbowls, were moveable objects, kept in a cupboard or storeroom and pulled out for occasional use. Typically made of copper or zinc, they were relatively light for ease of movement and, of course, waterproof. Galvanized steel is steel coated in a thin layer of zinc oxide which protects it from corrosion and oxidation and, as such, it will last a lifetime. Many of the tubs on the second-hand market date from as recently as the 1930s, when inside bathrooms were still a relative luxury, and have plenty of life left in them. Galvanized steel also has about an 85 per cent EOL (end-of-life) recovery rate, which means that it is worth recycling when you have finished with it.

Materials for a chilli pot

- Old bathtub
- Drill
- Recycled bricks, broken clay pots or stones
- Garden soil and compost
- Chilli plants

Method

Choose an old bathtub with a perished base or drill a few holes in the bottom to allow water to drain.

Fill the base with recycled bricks, pieces of broken pots and stones to leave an air pocket over the holes to prevent them getting clogged with soil.

Add a layer of soil or an upturned turf.

Fill the tub with garden compost up to a few centimetres below the rim.

Plant up with chillies, choosing a range with different levels of heat or just different colours.

Add some more compost as a mulch.

Water in and keep moist, especially as the chilli fruits are developing.

Project: Create a bathtub water feature

Even the smallest garden or balcony can attract some extra wildlife with a micro pond. This tiny body of water will soon attract birds to drink, and dragonflies may even visit if you are near a larger body of water. To help bees, leave a bit of moss on the edge so they have somewhere to land and drink, or add some stones to make a very shallow area and see who comes to visit. A pool of stagnant water is a favourite place for mosquitoes to lay their eggs, so to avoid a breeding ground for their larvae, make sure that the water does not stand still for too long by topping it up manually from a water butt every now and again, which will also inject some oxygen to help the plants grow and prevent the water going green.

Materials

- Old galvanized steel bathtub or bucket
- Oxygenating plants such as hornwort
- Marginal plants
- Gravel

Method

If you don't find a bathtub any watertight container will do. Choose a galvanized steel pot that does not have holes in it and that is at least 40cm/16in deep, since most aquatic plants need about 30cm/12in of water to grow in. **[1]**

Place a few oxygenating plants in the water. One non-invasive plant that you can throw in without a pot that just floats underwater is hornwort (*Ceratophyllum demersum*). It dies back in winter and sprouts new growth in spring. **[2]**

Add a couple of marginal or emergent plants: these usually live at the margins and emerge from below shallow water. Choose one with flowers to attract pollinators, such as water crowfoot (*Ranunculus aquatilis*) and slender club-rush (*Isolepis cernua*). These plants are usually sold in baskets rather than pots to allow water to circulate, and they can remain in these for a year or two. You should then take them out and divide them to keep them healthy. **[3]**

Top the compost in the baskets with a little gravel or a few stones to prevent it from floating away. If you are given a plant or are dividing plants from another pond as I did, you can just wrap the roots in some soil to feed them and cover the bundle with hessian tied at the top with string, before weighting it down with a brick. **[4]**

Place the plants in the pond where they will receive plenty of sunlight during the day. **[5]**

1

2

3

4

5

CHATEAU
MAUCAILLAUD
BORDEAUX SUPERIEUR
BONFILS
VINS & CHÂTEAUX
1870
www.bonfilswines.com

Project: Enjoy wine box herbs

If you are lucky enough to have wine delivered in wooden boxes, or have a friendly store nearby willing to donate some, you can use them to make excellent containers for herbs or cut-and-come-again salads. With beautiful images of chateaux and wine names on the side, they add the allure of the vineyard and a sense of occasion to any entertaining and growing space. These wooden boxes will also hang on to their embodied carbon until they break down, so this is a good way to prolong their life and create something beautiful and useful.

Materials

- Wooden wine box
- Old plastic bag (for indoor displays)
- Broken pots, stones or recycled packaging
- Garden compost
- Herb seeds or young plants

Method

Being porous already, the boxes don't need additional drainage holes, but if you are keeping them inside you will need to line them with a large recycled plastic bag or compost sack to stop them leaking.

Fill the base with broken crockery, tiles, stones, sticks or recycled packaging to add bulk and save on growing medium, then top up with garden compost.

Add a layer of fine sieved compost on the top so that you can sow seed directly into the boxes. I find them a perfect container for basil, which I use as a companion plant among my tomatoes. Alternatively, add young plants bought from a nursery, plant swap, or even the supermarket.

Fill with different herbs for a quick addition to the kitchen or barbecue – sage, oregano and rosemary are my perennial favourites. A wine box is also a good place to grow mint on its own (rather than in the borders), as this herb's roots love to run and can become invasive.

1

4

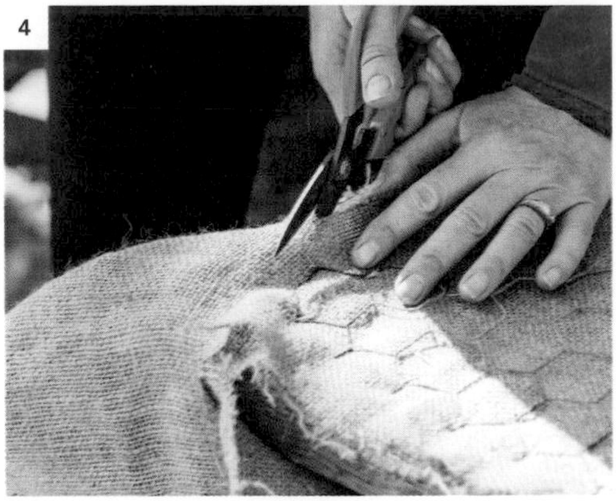

2

5

3

6

Project: Make a succulent frame

Ideal for a balcony or city terrace, or to create an unusual focal point in a larger garden, this succulent frame is the perfect low-maintenance solution to feeding pollinators without fuss. Houseleeks (*Sempervivum*) come in many colours and produce star-shaped flowers in pink, red or occasionally yellow. They are pollinated by bees and butterflies and their water-catching leaf rosettes also make them ideal spots for an insect to drink. To thrive, they just need good drainage and sunlight, and are often found squeezed into rocky crevasses high up in alpine conditions. Planting houseleeks by your front door is also meant to bring good luck.

Materials

- Reclaimed timbers
- Hammer and nails
- Staple gun
- Garden compost and grit or sand
- Moss from a damp area of lawn
- Chicken wire netting
- Hessian or recycled plastic backing
- Pencil or dibber
- *Sempervivum* plants either in 9cm/3½in pots or plug plants

Method

To make the frame, use reclaimed timbers of a size to suit. I built a rectangle, but any shape will work. Cut halfway into both ends of the long pieces, 5cm/2in from the end, and remove this notch of wood to leave an L shape. Then do the same for the shorter ends, so that when they are put together with the long pieces they overlap and can be fixed together with a nail. Lay your frame on the work surface, and fix chicken wire netting to the back of it with a staple gun. **[1]**

Turn the frame over and fill it with moss, pressing it down to create a layer about 1cm/½in thick over the wire mesh. Add a layer of gritty compost (mix three-parts compost with one-part grit or sand) on top of the moss and press down firmly, then top up until the frame is full. This prevents the soil from settling too much over time, which would loosen the plants. **[2]**

Attach a thick or double piece of hessian or recycled plastic to the front with a staple gun. This will then become the back. Attach fixings either side to hang the finished frame. **[3]**

Decide on a pattern or a random layout, and remove plants from their pots in turn, gently teasing off about two-thirds of the compost to leave a small amount around the roots. With a pencil or dibber, make holes and push the plants in through the wire mesh, leaving room for them to grow. **[5]**

Leave the frame flat for a couple of weeks to allow the roots to take hold and then hang it in a sunny spot. You can mist it occasionally if the weather is very dry and hot. **[6]**

1

4

2

5

6

3

Project: Create a willow dome

Willow is a thirsty plant that likes damp places, so do not plant it near drains or its clever roots will seek them out and may block them. It grows quickly and easily from cuttings, and all you have to do to get a stem to grow is to stick it in the ground. You can forage for willow with the permission of the landowners, or buy the stems, known as withies, from a specialist nursery. In late winter and early spring, when the willow is dormant and before it begins to shoot new leaves, you can buy kits for various projects that include withies of different sizes and colours.

Materials

- 20 thick willow withies or rods about 2.4m/8ft long
- 20 slightly slimmer withies about 2.4m/8ft long to create the lattice weave
- Optional thinner rods to create decorative elements such as windows
- String to secure the top and sides (optional)
- Strong sharp secateurs for pruning

NOTE: You can modify the design to suit the willow you have available.

Method

Mark out a circle about 1.8m/6ft in diameter using a 90cm/3ft piece of string tied to a stick in the middle of the circle. Cut the ends of the longest stems about 5cm/2in from the base to create a sharp point to make them easier to insert into the ground. Insert a 2.4m/8ft willow rod every 30cm/12in around the circle, leaving a space of about 90cm/3ft for the doorway. **[1]**

Cross opposite rods over at the centre and tie or weave them into one another at your chosen height (for example, 1.5 or 1.8m/5 or 6ft) to create the apex of the dome. You may need a second pair of hands to do this! **[2]**

Going back around the circle, insert slimmer rods next to the first, set at a 45-degree angle to create a lattice weave pattern. For fun, these could also be a different colour of willow. **[3]**

Either weave the second set of rods in and out of the first set, starting at the bottom and working your way up, leaving about 30cm/12in between squares, until you have created a chequerboard dome, or tie them together at the crossover points with garden string. **[4]**

Each year, as the willow grows, weave in the new growth and cut off any excess. If you have thinner rods left over at the beginning or after pruning you can weave a window or a pattern into the side too, or just free weave in where you want a bit of extra thickness. **[5]** You will soon have a wonderful secret hideaway for children – or adults. **[6]** If you want to keep the ground inside dry, add a layer of hessian and cover with a 10cm/4in layer of woodchip.

Project: Make a bird apartment

Songbirds are a beautiful addition to any garden, and they also work to keep the ecosystem in balance by eating excess insects before they become a pest. Here I have used an old wine box, but you could use any similar container. The key is to ensure that the entrance hole is about 4.5cm/1¾in in diameter, that the front can be removed to clean it out at the end of the season, and that it can be fixed to a structure high enough to protect the residents from most predators.

Materials

- Wooden box – this one is about 45cm/18in long and 30cm/12in wide inside
- Wooden vertical and horizontal dividers to divide the box into nests of 15 x 15cm/ 6 x 6in
- Drill with a 45mm/1¾in drill bit
- Dowels (optional) to provide perches
- Staples or tacks (10mm/½in size)

Method

To create the internal dividers, put notches in two pieces of wood the same length and three pieces the same width as the box, to slot into one another. Insert to divide the box into individual apartments of about 15 x 15cm/6 x 6in. I used one divider in the centre of the length of the box and two dividers the width of the box to make six nests in total. All dividers are the same depth as the box. [1]

Turn the box over and, with a 45mm/1¾in drill bit, make entrance holes in the centre of each area. [2]

Add dowels for perching by drilling holes just below the entrance holes and inserting them. These are not strictly necessary and should be omitted if you want to deter sparrows in the US, where they can outcompete purple martins for homes. [3]

Use a hinge or simply fasten the front closed with removable staples or nails, as shown here. This allows the front to be removed to allow the boxes to be cleaned in winter when the birds have flown to warmer climes.

Place as high as possible under the eaves of the house on any wall. About 6m/20ft high and north- or east-facing is ideal, to avoid over-heating in a south-facing suntrap. The box does not need to be sheltered from rain. [4]

If you want to help further, leave out twiggy sticks, moss or bits of wool to provide nesting material, and leave muddy patches of ground, if you have them, which are ideal for nest-building birds. Swifts and house martins live on insects in flight, so by avoiding pesticides and providing plants and some water to feed the insects, we are also feeding the birds.

Les TOURELLES de LONGUEVIL
LLES de LONGUEVILLE
CRD
2011
1

TOURELLES de
2

3

4

Plants really bring a sense of magic into the garden.

Cultivating place

The real magic of gardens begins once you've decided on the structure and can start to bring in plants. It's tempting to head straight to a nursery or garden centre, but before you do that a moment of pause and planning can save a lot of money and a few mistakes.

First decide how much time and budget you have available. If possible, you should start with the structure: trees, if you can fit them, then hedges, shrubs and climbers. After that, consider herbaceous perennials – plants that come back each year from their energy storage below ground – and, finally, annuals, which are plants that grow quickly from seed, flower and set seed for the following year before they die. Your choices will range from the most sustainable and budget-conscious extreme of planting an acorn in your garden and waiting thirty years for a beautiful tall oak tree, to the least sustainable option of shipping in a fully grown oak from abroad and craning it into place. Most of us would choose somewhere in between.

While there may be immediate gratification in an instant garden, it is less joyful when you know the cost to the planet in terms of fossil fuel usage. The wait can be balanced with the deep joy of planning for the very long term. Planting something that may outlast us and will carry on caring for the generations to come gives an amazing feeling of connection with both the past and the future.

Choose plants that will thrive in your garden to minimize added care.

Plants for purpose and place

We have looked at how to assess the different areas of your garden (*see page 36*), so let's now translate that to what to plant where. Armed with a list of the types of plants you want, from trees and shrubs to climbers, perennials and low-growing ground cover, and having checked their site requirements, such as their need for sun or shade and damp or dry soil, you can now search in the knowledge that your investments should flourish.

The first question is, what do you want the plant for, besides its superficial beauty? For example, if you need a climber to cover a shady wall, you may also want it to feed bees, provide habitat for nesting birds, and be evergreen. You may not be able to accommodate all the points on your wish list in one plant, but with these parameters an ivy will tick all the boxes, as will a climbing hydrangea, except for the evergreen option, since it loses its leaves in winter.

You can turn the question around if you find a plant you love and want to know where to put it: by looking up a plant's origins, you will begin to understand its needs. If it comes from a rocky hillside in the Alps, you can imagine that it will endure cold and like well-drained soil, but it won't like lowland clay where its roots will sit in wet soil all winter. Or, if it grows naturally in woodland, it will suit a shady spot in summer, and if it is a Mediterranean plant, it will be happy in a fair amount of sun and will tolerate winter cold but not wet soil. If it comes from the tropics and you live in a cool country, such as the UK or the US in USDA zone 7 and below, you will have to protect it or bring it inside in winter, so make sure you have space. Once you know your own soil type you will be able to match it with plants that are most suited to it (*see pages 56–8*). If you can select plants that would choose to grow in your garden naturally, you will be going with the energetic flow, rather than battling against nature by trying to change your soil with additives or chemicals.

Plant fairs are a great source of plants, knowledge and friends!

Sourcing plants

Having decided on what to plant, the next question is where to buy them? The internet allows you easy access to independent nurseries staffed by passionate and generous gardeners who will give advice and share in your search for the perfect dry-shade-loving shrub or sun-loving annual.

Check whether their seeds are sustainably sourced and their plants are grown in peat-free compost. You may even find that packaging and boxes are compostable, rainwater is used to water plants, and renewable energy heats the glasshouses. If they are local or can use shared transport to deliver, you have hit the jackpot. All that is left to choose is what to buy.

Plant fairs are other excellent sources of plants and knowledge. Groups of plantspeople gather to trade ideas and stories as well as to sell plants, and some like those at Great Dixter in the UK and the Madoo Conservancy near New York include expert talks to inform and inspire visitors.

Plant swaps are another great way to make gardens and friends. If you like a plant in a neighbour's garden, you could ask for a cutting (do always ask) or why not offer one from your plot to someone else. Seed packets often have too many for one year's sowing, too, so you could organize a seed swap, either locally or by post.

Independent nurseries can offer a wealth of information and often grow more unusual plants.

Choosing plants

The importance of trees

We could dedicate a whole book to trees, and with good reason. Trees lock up carbon, reduce the volumes of greenhouse gases in the atmosphere and also clean the air in our cities by filtering dust and particles of pollution. They provide roosting and nesting spots for birds, and habitats for insects and bats in crevices and cracks and under their bark. When we use them for building and making furniture, trees continue to hold on to the carbon they have extracted from the air, until they gradually release it together with nutrients as they decay back into the soil. They are the ultimate renewable energy source.

They also regulate the air temperature. In our cities trees can reduce the air temperature by as much as 2°C/35.6°F, which in some cases can be the difference between life and death. If we think about where to plant trees around our houses we can reduce our reliance on air conditioning and heating systems by using them for shade in the summer and to provide shelter from cold winds in the winter. For example, a deciduous tree, which loses its leaves in winter, can be planted on the south side of the house, where it will cast shade during the hottest months when it is in leaf, while in winter the sun will penetrate its bare branches to send warmth through the windows.

On a larger site, trees provide shelter from the wind by sending it up and over in an arc, filtering the wind on the way to reduce its impact. In the same place a wall would create an eddy on the lee side, and two walls might funnel the wind, speeding it up to create an uncomfortably draughty corridor. Trees work as an anchor in the garden, too, not only providing height and shade but also creating a steady ecosystem in the soil. Their roots reach out and stabilize the soil, and draw to them a whole fungal system of microorganisms under the ground, which work together to exchange sugars and nutrients between plants. They even send messages and can warn of an attack by insects to stimulate trees to release bitter-tasting chemicals in their leaves to ward off the attackers.

So, however small your garden or allotment, if you have room for just one tree, you'll be doing the world a favour.

Trees provide flowers, food, clean air and shelter, as well as sequestering carbon and anchoring the soil. They are vital to our survival.

A natural log hive for wild honey bees.

Crab apples feed the birds in midwinter.

Trees for bees and wildlife

As well as providing habitat, shade and carbon capture, trees are a great addition to a sustainable garden because they provide lots of different types of food for us and for wildlife. If you have room for some espaliered fruit trees, you can grow pears, apples, plums, or even peaches if you can give them enough heat or grow them against a warm wall. Or grow a nut plat, which is a traditional way to grow cobnuts or hazelnuts: it's laid out in a quincunx or staggered row to maximize the use of space and aid cross-fertilization.

Trees also feed many different insect species and all demand relatively little work from us. One of the main food sources from trees is the nectar in their flowers, which will fill a tree with the buzzing of bees when the blossom is out, and, if you have a lime tree (*Tilia*), a sweet scent too. The crab apple tree is another perfect example: its blossom feeds insects in spring, and in the depths of winter, when the ground is covered in snow and there is little to eat, its tiny apples provide nourishment for birds. With care you can choose trees to flower and provide forage for bees year-round in many climates.

Choose trees to flower and fruit through the year for a garden full of birds and bees.

BLOSSOM CALENDAR

1. *Salix alba* (late winter to late spring)
2. *Prunus avium* (early to mid-spring)
3. *Amelanchier lamarckii* (early to late spring)
4. *Malus domestica* (early to late spring)
5. *Aesculus hippocastanum* (mid- to late spring)
6. *Liriodendron tulipifera* (early to midsummer)
7. *Tilia cordata* (midsummer)
8. *Koelreuteria paniculata* (mid- to late summer)
9. *Eucryphia glutinosa* (late summer to early autumn)
10. *Arbutus unedo* (mid-autumn to early winter)

Sustainable shrubs and climbers

After trees, shrubs and climbers form the backbone of the sustainable garden. When chosen carefully they will provide structure, texture, colour and scent all year round, as well as places to live, nest and forage for lots of wildlife, both above and below the soil.

When we invite a person to a party, we don't just think about what they look like and the same is true of plants. By considering their properties and functions, as well as their appearance, we can create a mini ecosystem that needs minimal additions from us throughout the year and is almost self-sustaining. By planting in layers (*see page 120*), you can protect the soil, lock up carbon and prevent water run-off, as well as creating a beautiful mini ecosystem. Shrubs also create the structural layer in the garden, working together with smaller plants to provide a year-round covering that protects and can also feed the soil.

Certain shrubs and climbers work together with fungi to fix the plant nutrient nitrogen from the air into nodules in their roots, which minimizes the need to bring in additional fertilizers. These include members of the pea and bean family, from French beans to sweet peas (*Lathyrus odoratus*), plus shrubs such as oleaster (*Elaeagnus angustifolia*) or red chokeberry (*Aronia arbutifolia*) in the US where the oleaster can be invasive, and sea buckthorn (*Hippophae rhamnoides*). Some trees, including birches (*Betula*), black locust (*Robinia pseudoacacia*) and Italian alder (*Alnus cordata*) are also able to perform this trick.

Between 1960 and 1990, global use of synthetic nitrogen fertilizer increased more than sevenfold, while phosphorus use more than tripled. Nitrogen run-off from fertilizers, car emissions and septic tanks ends up polluting water courses and causing eutrophication, indicated by an algal bloom that produces toxins harmful to plants, animals and humans. By creating leaf mulch and compost we are already saving the nitrogen stored in plants' leaves (*see pages 59–65*) and if we can include some of these nitrogen-fixers as well, we will be adding even more nutrients to the soil bank.

In addition, by planting to attract beneficial predators we can prevent infestations of plant-devouring insects and so avoid using pesticides, which are currently threatening the bees and other pollinators we rely on to pollinate our food.

So how do we choose shrubs and climbers? When planning what to plant, we want to make the least change for the greatest effect, so begin by seeing what is growing in your garden already and how you can work with that. If you have a fence covered in ivy or a wall with moss, think twice before stripping it all off, since moss is one of the best air purifiers, and mature ivy flowers are a favourite forage for bees, while the plant as a whole offers a great place for birds to nest. So, keep a bit of the beneficial stuff, clip it back (but not during nesting season) if you are tidy minded, and then add to it.

Next, choose the bulk of your shrubs and climbers from communities of plants that grow nearby and do well, either in neighbours' gardens or even in neglected corners in a town. Some shrubs such as buddleia are pioneer plants that colonize cracks in walls and fill them with flowers for butterflies. These tough plants will do well in any garden and may just need to be kept in check in warm humid environments where they can do too well and get out of hand.

Once you have the shrub layer, add climbers such as roses, clematis or honeysuckle for additional height and a longer flowering/nectar period. You can tie them to wires or trellis fixed to a fence or just let them scramble among other plants as they would in the wild.

Ivy is a brilliant climber for bee forage.

Honeysuckle will provide nectar and colour.

Prunus serrula looks lovely, while providing structure for the garden and shelter for the birds.

Layered planting takes care of the soil and provides interest throughout the seasons.

Sustainable perennials

Below the tree and shrub layer is where the most intensive gardening often happens, with some gardeners historically changing the contents of the perennial and annual plants in their borders every season. However, with careful planning, this area can be less work-intensive, and you can create a succession of colour and interest without disturbing the soil too much or using artificial fertilizers.

To begin, think of the roles and functions of each plant and how they will look in each season. It is good to have a mix of shapes and sizes, not just for beauty but also to provide a mix for foraging insects. The flat heads of umbels such as cow parsley (*Anthriscus sylvestris*), with their many flowers packed together like an umbrella, are the perfect landing pads for dragonflies, while the spires of foxgloves have a doormat on every flower, designed to guide a bumblebee inside to drink nectar, and rub their fur with pollen in the process.

To be sustainable a perennial also needs to be happy in your soil and aspect so it will thrive without too much intervention, but not become invasive to the detriment of a balanced ecosystem. The butterfly bush (buddleia), for example, is a plant that colonizes cracks in pavements and old buildings in the UK, and it�s a great nectar source for butterflies, but in the US its copious seed dispersal means it can become over-dominant in the landscape, to the point of being banned in some states. The same is true of purple loosestrife (*Lythrum salicaria*), which is a fantastic pollinator plant and a strong doer in the UK but a banned invasive in the US.

A sustainable plant is compatible with the plants around it; the mix of plants becomes as important as the individual. Large-leaved tall plants will stifle sun-loving plants below but be perfect partners above shade-lovers, while alpine plants that need plenty of space to spread will be squeezed out of a crowded fertile border but thrive on their own in a rocky corner. We should also consider the depth and spread of roots below the ground. To allow different plants access to different layers of water and nutrients in the soil, you can plant a mixture of grasses with shallow roots, shrubs with mid-layer roots and deep-rooted perennials such as *Baptisia australis* with long taproots. Think of your planting as a layer cake above and below ground. Of course, the fun of gardening is learning through trial and error, by observing what thrives and what fades out through the seasons. By creating or following a chart like the one overleaf, you can build up your own palette of sustainable plants that look good and work well together.

Creating a perennial mix

- Choose three perennials and one grass to suit the soil, amount of sun and shade, and amount of likely rain your garden receives.
- Check online or in a good reference book how tall the plants will reach and how much they will spread after three years. This may vary between 20cm-150cm/8in–5ft depending on the plant and the conditions.
- Check where the ideal conditions for the plants would be in your garden.
- Note which animals they will feed and any other benefits or drawbacks they have.

Start to fill in the chart below and you will see whether you have a good spread of roles, functions, sizes, shapes and colours with performance through the season. If you find big gaps in seasonality or too many plants performing the same role, you can adjust your planned planting accordingly. Once you have this for one mix, you can repeat it to create three mixes that work well together.

LAYERED PLANTING

AESTHETIC INTEREST

Category	Plant	Evergreen	Spring	Summer	Autumn	Winter
TREES & SHRUBS	Hornbeam	N	Flower	Foliage	Foliage Fruit	Foliage
SHRUBS & STRUCTURE	Euphorbia	Y	Foliage Flower	Foliage Flower	Foliage	Foliage
	Sedum	Y	Foliage	Foliage Flower	Foliage	Foliage
	Calamagrostis x acutiflora 'Karl Foerster'	Y	Foliage	Foliage Flower	Foliage Flower	Foliage
SPOT PLANTS	Kniphofia	Y	Foliage	Foliage Flower	Foliage Flower	Foliage
	Allium	N	Foliage	Foliage Flower	-	Foliage
	Ammi majus	N	Foliage	Foliage Flower	-	Foliage
EMERGENT HERBACEOUS	Crocosmia	N	Foliage	Foliage Flower	Foliage Flower	Foliage
	Helenium	N	Foliage	Foliage Flower	Foliage Flower	Foliage
GROUND COVER	Geranium	Semi	Foliage	Foliage Flower	Foliage	-
	Sanguisorba	N	Foliage	Foliage Flower	Foliage Flower	-
	Alchemilla	N	Foliage	Foliage Flower	Foliage Flower	-

ECOLOGICAL				OTHER USES	
Nitrogen fix.	Ecosystem support	Root structure	Resilience	Edible	Medicinal
Y Nitrogen fixing nodes	High	Fibrous – shallow & wide spreading	●	Some parts	Tonic – fatigue relief. Leaves – stop bleeding & heal wounds
-	Good soil covering	Taproot – shallow & narrow	●	Toxic	-
-	Year-round ground cover	Fibrous & creeping – shallow roots	●	Mostly	Relief from pain & inflammation
-	Good for grass dwelling species	Shallow & fiborous	●	-	-
-	Attracts pollinators – butterflies & moths	Shallow bulbs	•	Non-toxic	-
-	Good for companion planting with crops due to aroma	Shallow bulbs	●	Mostly, in moderation	Lowers cholesterol, blood pressure & helps to prevent cancer & cardiovascular disease
-	Attracts pollinators	Taproot – medium depth	•	Can be toxic	Has been known to treat chronic ulcers
-	Hummingbirds love the tubular flowers	Corn root system – shallow	●	Corn root system	East Africans treat malaria with a beverage made from its leaf sap and corms and a rheumatism-treating drink from its roots
-	Attracts pollinators	Shallow & wide-spreading	●		'Sneezeweed' – inhaled to cause sneezing
-	Excellent year-round ground cover	Once established can create a drought-resistant soil ecosystem	●	Yes	Immunity boosting qualities. Used in relation to menstrual problems
-	Important food for the larvae of some butterflies	Shallow, fibrous root system, extensively used for erosion control	●	Leaves can be used in salad - said to taste like cucumber	It is said to cool the blood, stop bleeding, clear heat and heal wounds
-	Nectar-rich flowers which attract pollinating insects	Shallow root system – rhizomes	●	Yes. Usually prescribed as a tea made from the flowers	Anti-inflammatory & astringent. Good for regulating menstruation

Project: Plant a layered border

Armed with your plant mixes (as explained on pages 117–19), and with a list of suitable ground-cover plants and shrubs for your garden, you can create a layered border that will become its own mini ecosystem, providing support and nourishment for itself and the creatures that come to live in it.

Materials & tools

- Cardboard and compost to create a no-dig border
- Spade or trowel
- I tree per 6 metres/ 20 feet (optional)
- About 9 ground-cover plants per square metre/11 square feet in 9cm/3½in pots
- 3 taller perennials per square metre/11 square feet
- 1 grass per square metre/11 square feet
- 1 shrub for every 2 metres/6 feet
- Composted bark or garden compost to mulch

Method

Prepare the border as per the no-dig vegetable bed (*see pages 138–9*).

If you have space, plant one tree for every 6 metres/20 feet, choosing trees that allow through plenty of light through their canopies.

Start with ground-cover plants suitable for your soil and the amount of sun and rain available, and plant them to cover roughly 50 per cent of the ground.

Add one shrub for every 2 metres/6 feet, mixing evergreens and flowering and fruiting shrubs to attract pollinators and feed animals, and maybe yourself. Aim for at least one shrub to be nitrogen-fixing (*see page 114*).

Add in one or two mixes of perennials, ideally with three perennials and one grass in each mix, to flower in succession through the seasons. Choose a variety of heights and flower shapes, from umbrella-shaped to spires and round daisies, to aid pollination as well as to look beautiful.

If you have gaps while the shrubs fill out, add in annuals for a splash of instant colour, to feed the bees and to protect the soil. They will seed about where they are happy and create a dynamic feel to the planting as it changes. Where there is any space remaining, cover with a mulch of garden compost or fine composted bark.

Water your plants in and keep their roots moist for the first three months until their roots are fully established (stick your finger deep into the soil to check for moisture before getting out the watering can). After that, they should only need watering in severe drought.

Plants for free

There are few things as satisfying and exciting as watching the tiny green shoots of new seedlings emerge from the soil. It's witnessing a miracle every single time. The process can be even more satisfying if the seeds are saved from your own plants, and this is incredibly easy to do.

Simply cut some of your favourite flower heads in the autumn and leave them to dry by hanging them up or setting them on paper in a warm place indoors. When dry, shake the heads gently on to the paper or into an envelope to release the seeds.

Store the seeds in a cool dry place over winter and then sow them in spring. I tend to wait until after the first new moon in early spring, and sow in shallow trays of peat-free compost. Recycled cardboard food trays are perfect for most seedlings, since they absorb water and then gradually break down in the compost later. For deeper-rooted legumes, such as peas, sweet peas and beans, that don't like the root disturbance of being separated and repotted at seedling stage (known as 'pricking out') you can sow individual seeds into recycled lavatory rolls and plant them straight out when they are large enough and after the frosts have passed.

If you haven't had time to collect your own seed or want more varieties, reading seed catalogues is one of winter's most pleasurable daydreams so just choose from a good organic independent seed supplier, and then organize a seed swap among friends or neighbours to redistribute any excesses. There are nearly always too many seeds in one packet so it's a great way to share plants and get to know people.

Once your borders are up and running you will find some perennial plants may become too abundant, and you can thin these by digging them up and splitting the clumps into four with a sharp spade in autumn or early spring. Either replant the clumps or give the extras to a friend who will hopefully offer you some plants or seeds in return. Gardeners are a generous lot.

Once you start propagating plants your main issue will be giving away the extras.

1

2

3

4

5

6

Project: Grow plants from cuttings

If there is a plant you love and would like more of, or if you have one that is tender and may not make it through a cold winter, it's a good idea to propagate by taking cuttings. Once you realize how simple this is, it can become addictive. If you are visiting other gardens do always ask before taking cuttings, and make sure you have a watertight compostable bag, paper towel and water in your bag to keep the stems alive until you get home.

Materials

- Clean sharp knife, secateurs or scissors
- Small 9cm/3½in pots
- Peat-free seed compost
- Pencil or dibber
- Rooting powder (optional)
- Transparent compostable plastic bags
- Rubber bands

Method

Fill a pot with peat-free seed compost, firming it gently to remove any large air gaps. Cut the stem just below a leaf node (where the leaf joins the stem or where you can feel a bump on the stem), which is where the plant stores its growth hormones. It will now use these to grow roots instead of leaves. **[1]**

Strip off any lower leaves which would rot under the soil. Also remove or cut in half any larger leaves and pinch out flower buds at the top. You want all the energy to be focused on growing roots, not supporting big leaves or flowers. **[2–3]** You can dip the cutting in rooting powder to increase the chance of success. **[4]**

Use a dibber or pencil to make a hole and pop the cutting in deep enough to support the whole stem, then firm it in gently either side of the stem. You can fit several cuttings in one pot. **[5–6]**

Cover the pot with a transparent compostable plastic bag to keep the moisture in and secure it with a rubber band (ideally one dropped by the post person, to prevent it harming wildlife).

Once you can see new growth, remove the bag. Fill another set of pots with fresh compost. Gently tease out each of your cuttings, easing the roots out with the dibber or a blunt pencil and holding it by the leaves. Then use the dibber to make a hole and insert the cutting into its own pot.

Leave it to grow on until the roots fill the pot (have a look by gently tipping it into your hand sideways, while cupping the stem to stop it all falling out).

Plant outside in the garden in autumn or spring for hardy plants, and after all risk of frost has passed for tender types.

Project: Saving seed

Collecting seeds is a beautiful way to celebrate the end of the growing cycle and the knowledge that plants will grow again next season. It is like throwing a line of hope across the darker winter months to the following spring. Many annuals, such as foxgloves, love-in-the-mist, sunflowers, poppies and honesty, are very easy to grow from seed and will seed about by themselves if they like where they are. By collecting seeds we can encourage them to seed elsewhere in the garden, or give seeds to friends, or take them to seed swaps. If a flower has been cross-pollinated by insects with a slightly different variety their seeds may not 'come true', i.e. be exactly like the parent, which adds to the variety in the garden. If the seed has been bred by a company and is what is known as an F1 Hybrid, it will not come true to seed. Cherry tomatoes are often an example of this, so their offspring will be slightly different, but may be all the more fun to try.

Materials

- Seed heads
- Scissors or secateurs
- Paper bag
- Seed trays
- Peat-free seed compost

Method

To collect seed, chose a dry day and find nicely mature seed pods that have not decayed, gone mouldy or been eaten by insects. Cut off the whole stem of the plant you would like to propagate with the seed head intact.

Hang the stem up to dry in a cool, dry place or place upside down in a paper bag to allow seeds to drop. For pea and bean seeds, you can place them on dry rice to remove excess moisture and prevent rotting.

Sow the seed in autumn or spring by gently opening the seed pods and sowing directly into the garden. Alternatively, sow them into seed trays and cover with a fine sprinkling of compost. Keep them slightly moist in a greenhouse or on a windowsill until they germinate.

For tomatoes and fleshy seeds, see page 153.

Project: Plant a gravel garden

A gravel garden can help to minimize maintenance and offers a home for a range of plants that don't like 'wet feet', which can be a problem in gardens with heavy clay soil (see pages 56–8). It's a great way to use a front garden that has to double-up as parking, especially if you select plants that will creep across the gravel and tolerate the occasional pressure of a car's tyres driving over them.

The gravel is essentially an inert deep mulch, which protects the soil below, helps to prevent evaporation and run-off and locks up carbon. I buy my plants in small 9cm/3½in pots, which have just enough root to get away quickly but avoid the extra transport fuel, soil and pot material of a larger pot and plant. The gravel supports the plants as they get their roots down and protects their leaves and stems from any excess water sitting in the soil – especially useful in heavy clay which holds water and rots more sensitive roots in winter. It also allows plants to self-seed naturally.

Gravel extraction can have its own issues, so check that yours is locally sourced and has not been strip-mined from the seabed, which kills marine life. In the garden shown here, I used plants that would be found naturally in a prairie setting, since it is on very heavy clay with compacted soil. Prairie plants have a mix of long taproots and shorter fibrous roots and will tolerate extremes of drought and wet, making them a good choice for these conditions. In a sandy soil which retains less moisture I would use a Mediterranean plant palette.

Materials

- Additional compost if you garden on clay soil
- Any sustainably sourced gravel that is free from cement and other contaminants
- Selection of plants

Method

To prepare the bed, add home-made well-rotted compost if your soil is very heavy, and then lightly roll it to get a firm but not compacted surface. Add a 15cm/6in layer of gravel – we use South Cerney gravel for its lovely flat, round shape, which compacts slightly so that it is easy to walk on without that swimming-through-treacle feeling, and yet allows enough movement to let plants come up through it. For most drought-tolerant plants you can use a 20cm/8in layer and plant directly into the gravel; for hungrier shrubs like roses, use less gravel so that it is just a mulch and the roots are totally in the soil.

You will need to water during the first season until the plants are well-established to ensure that the small root systems do not dry out. Apart from that, you will only need to deadhead if you want a longer flowering period, or just cut the plants back in late winter before the new flower shoots have begun to grow.

Project: Create a wildflower border

Sown at the right time of year on to subsoil, wildflower seed can establish very quickly, but if you are in a hurry, there are companies that sell wildflower turf pre-sown with a variety of mixtures suitable for different situations, depending on your levels of shade, soil moisture and acidity. These are very simple to just roll out on top of tilled soil that has had any invasive plants such as docks and thistles removed. It can be cut to shape (like regular lawn turf, making sure that there are no small sections at the edges that will dry out) and then watered in.

Materials

- Wildflower seed or plastic-free turf
- Yellow rattle seed (*Rhinanthus minor*)
- Mower
- Spring bulbs (optional)
- Sharp spade

Method

Start with one strip of lawn, perhaps the width of the mower, or larger if you prefer, and leave it unmown. Add in some bulbs, such as camassia, narcissus and crocus in the autumn and allow the grasses and flowers to bloom in spring and summer.

Alternatively, scalp a strip of lawn by mowing very closely to expose the soil in places. Into this, sow wildflower seeds and some yellow rattle, which will reduce the vigour of the grasses and allow the flowers to establish.

A third method is to strip back all of the turf by cutting into it with a spade and then using the spade to slice through the roots underneath it, releasing slabs which you can then use to create a hügelkultur mound (*see pages 140–41*). Then sow wildflower seed on to the subsoil soil you have uncovered. Subsoil contains fewer nutrients than topsoil, and allows the wildflowers to thrive while the hungrier grasses are kept in check.

A wildflower border, whether it is made from a lawn alone or sown from seed, only needs scything or strimming once in late summer or early autumn, after it has gone to seed. Leave the cuttings for a couple of days for the seeds to drop down into the sward.

Remove the hay before it breaks down to avoid enriching the sward with nitrogen, which will favour grasses over flowers. If you have a friend or local garden with a native wildflower meadow, late summer is also the time to ask them for their hay. When strewn on your ground for a few days it will add to the species in your meadow or allow you to start from scratch on a patch of subsoil or lawn for free. This is seed swapping at its best.

Pottering in the vegetable garden is one of my favourite ways to feel connected.

The vegetable garden

Considering the effort and cost of producing your own vegetables, you may ask if it is worth the trouble, or whether you should leave it to the professional growers who can produce and deliver with all the associated economies of scale. However, when we think about the importance of eating simple, unprocessed organic food for our health, and consider that an estimated 33 per cent of all food produced is lost or wasted, making up more than 50 per cent of global landfill waste, the argument swings in favour of both knowing where our food comes from, and understanding the process of growing it. Studies also show that eliminating global food waste would have the same effect as taking one in four cars off the road, saving 4.4 million tonnes/ 4.9 million tons of carbon dioxide pollution a year.

There is no better way to reduce this pollution than to grow some food yourself. We are unlikely to throw away a single leaf from a bowl full of salad leaves that we have sown, grown, picked and washed to present at a meal. Add to that the joy of tasting truly fresh vegetables and understanding the plant properties that create the goodness in each mouthful. Then include the mix foraged greens and flowers from the garden and hedgerows that are nature's gifts and we can already feel so much closer to the earth. Growing food is a wonderful way to stake our place in the universal system that we are all part of.

Like everything in the sustainable garden, though, only grow food at your own comfort level, without judgment or creating a rod for your back. You could start with a pot of basil on the windowsill or a tub of tomatoes and herbs by the back door – once you taste home-grown tomatoes, you may be hooked. The following pages have some ideas for produce you might like to try; just pick and choose what you are drawn to, there are no hard rules. Gardening is about learning as you go.

Crops to grow

When choosing what to grow, the first question has to be, what do you like to eat? If you love salads, then some cut-and-come-again leaves grown in an old wooden wine box are a joy to add to every lunch bowl. Sprinkle on a few flowers or petals to create a bowl of beauty.

Some foods taste totally different when harvested straight from the plant, and I would always add these to your list. Peas rarely make it into the house from our vegetable garden; they taste so good raw and lose their edge to become starchy after just a few hours. Tending fresh tomatoes will make your hands smell divine and fill you with feel-good endorphins, before you even add the taste of a few popped into your mouth in the greenhouse. The flavour of just-pulled carrots is another one that is impossible to find in shop-bought roots, and the fresh tops make great pesto. In my view, organic potatoes taste as good from my farm drop, but digging for them is like uncovering buried treasure, a thrill for children and adults too. You will find your own favourites, of course, and in all cases just putting your hands in the soil will increase endorphins and boost your immune system.

Once you have a list of what you want to grow to eat, check how long they take to mature, how much space each plant needs and when it will

need harvesting. This will help you to plan the layout of your beds and planting timings to avoid both sowing too much at once and harvest gluts. A chart or even a spreadsheet can do wonders to map out tasks and prevent feeling overwhelmed. Once you have harvested your spring and summer vegetables such as salads and early potatoes, you can plant for an autumn or winter harvest of leeks or calabrese. Use a home-made or organic peat-free compost as a mulch to feed the soil after harvesting a nutrient-hungry crop and to cover the beds after planting, since healthy soil means healthier plants and more nutrients in every mouthful.

Growing under glass

Since the royal physicians pioneered growing under glass in around AD 30 to feed the Roman Emperor Tiberius healthy, year-round cucumbers, we have tried to harness the sun's energy through glass. In cooler countries with shorter growing seasons and variable day lengths, it can have a huge impact on what we can grow and definitely adds to our ability to be more sustainable.

As ever, there are decisions to be made, such as whether to heat a greenhouse in winter, which will impact its sustainability. If you are able to use

Choose vegetables you love to eat, whether carrots, peas or leeks.

green electricity this may be an easier choice, but if you rely on oil or gas, it is usually better to stick to growing plants that can gain a big boost with no extra heat. To protect plants in an unheated greenhouse, try using hessian or wool fleece to cover seedlings during a cold snap and cardboard as insulation under trays if you have a concrete base. A watering can filled with water can also help mitigate a light late frost: it will be warmed by the sun during the day and when placed among your plants it increases the air temperature slightly during the night. Hessian and fleece can be used in summer to provide shade when needed too.

Siting your glasshouse in the lee of a deciduous tree to allow sun through in the winter months and provide shade in the summer will also make it more temperate, and when setting up a greenhouse think of building by function, using upcycled staging, windows and doors wherever possible.

Laying out the veg plot

As sustainable gardeners we are always trying to create the perfect microclimate for our plants to thrive. A classic vegetable garden is best placed where it will receive sun through the day in spring and autumn, and a little shade in midsummer from deciduous trees. In a cold climate, avoid frost pockets, which is where cold air gets trapped at the bottom of a hill or on the north side of a hedge or wall where the wind can't blow it away. In a hot climate, use trees to filter drying winds. Walls that receive the midday or afternoon sun will retain heat and can be useful for ripening fruit, especially when painted white to reflect the sun on to the plants.

It's also important to have a water source nearby and while a glasshouse or cold frame will extend your growing season, a windowsill indoors works just as well for a small space.

A traditional vegetable plot has raised beds with paths in between for ease of access and to avoid soil compaction from treading on the beds. If you have a small garden or little time, one raised bed may be plenty for both food and fun. To create this micro garden, you will need an edging material such as woven hazel hurdles, old scaffold boards or reclaimed bricks for the sides. If using softwood you could treat it with a non-toxic oil, but avoid sleepers impregnated with creosote, which is toxic and may leach out.

Apply a mulch over the beds. I use an angular chipping gravel, but wood chip or compost over cardboard would also work well.

You will then need a mix of vegetables grown from seed or bought as plugs, such as:

- Dwarf climbing beans
- 3 cabbages
- Mustard greens or other leaves
- 1 courgette
- Nasturtium and flat-leaf parsley

Plant the beans towards the back of one bed and make tepees out of twigs or canes tied at the top to support their stems; the beans will scramble up as they grow. Dwarf climbers will only grow to about 60cm/2ft, but other beans will need 1.8m/6ft supports or a trellis fixed to a fence. Arrange the cabbages below the beans, with the mustard greens or other leaves in between.

In a second bed, grow the courgette under-planted with nasturtium and parsley. Both the courgette and nasturtium flowers are edible and look lovely in salads.

The beans fix nitrogen and the mustard greens may deter cabbage white butterfly larvae, but be vigilant and pick off any caterpillars you see to put on the bird table.

Hazel rods make good bean supports underplanted with leafy vegetables.

Project: Adopt the no-dig approach

No-dig borders are much less prone to compaction than traditional borders because the soil structure is not damaged by digging. Without the additional height and edging material of raised beds to divide them, borders can become fluid and the vegetable and ornamental garden can merge into one another without barriers.

There is no need to stick to the rigid rows of the traditional vegetable garden, and the principles of choosing plants for their function and planting in layers still hold. For example, you might choose a perennial cavolo nero cabbage, kale or chives, and plant annuals such as carrots in between. The kale roots will anchor the soil and provide some shelter for the carrot seedlings and as you pick the perennial's leaves, the smaller plants will be exposed to more light too. Cavolo nero and kale will regrow for three or four seasons before they need replacing. As well as providing tasty stalks and flowers for your salads, the smell of the chives may help to deter pests like carrot fly, while the flowers also feed bees.

Materials

- Sheets of cardboard – flattened delivery boxes are ideal
- Home-made compost or peat-free garden waste
- Well-rotted manure

Method

Lay the cardboard over the soil in the area you plan to plant your crops, overlapping the edges to ensure there are no gaps. **[1]**

Cover with a thick layer of compost, about 15cm/6in deep. **[2]**

Allow the compost to settle through a couple of periods of rain and then plant seedlings straight into the compost. **[3]**

As the plants send their roots down, they will grow through the cardboard, which will have begun to break down.

Weed little and often to take out self-seeders and any unwanted perennials such as nettles or docks. A quick hoe when plants are tiny creates minimal root disturbance for your crops. (Really deep-rooted perennials such as bindweed may need to be covered with recycled black plastic for a year prior to planting to exclude light and kill the roots, which can be more than 1.8m/6ft long.)

Feed the soil after harvesting with a new layer of compost and you will be rewarded with lots of worms, beneficial bacteria and fungi to create a healthy ecosystem. Composting is an inexact art, but you can tell that compost is healthy if it is dark brown, moist but not wet and smells delicious. **[4]**

1

2

3

4

Project: Make a hügelkultur mound

Hügelkultur or 'mound culture' is a clever extension of the no-dig system which increases the surface area of a border by building it up into a mound. It also adds to the soil's fertility below ground as the logs, twigs and compost used to make it gradually break down over time and release their nutrients to be used by the plants' roots. The decaying biomass inside the mound extends the productive time of these beds through the year, especially in hot places where the logs and twigs behave like a sponge, reducing the irrigation needed, and in cold places where it will generate some heat throughout the colder seasons.

Over time, the mound will settle and reduce in height, so if that is an issue, you can follow the same process of layering with logs, sticks and leafmould plus a little topsoil in raised beds and plant them up with a mix of vegetables (*see pages 134–6*). Or, on a hilly slope, you can use hügelkultur to create terraces. On a flat area, they can be planted with taller shrubs or hedging to create windbreaks, making a useful natural addition to larger-site planning and microclimate management.

Materials

- Logs of a variety of sizes to suit your space
- Sticks and twigs
- Leafmould or compost
- Straw or composted bark
- Topsoil
- Seeds or plug plants or larger plants

Method

Dig a shallow trench [1] and remove any turf and topsoil. [2]

Line up large logs as shown (the older the logs the better as they will be more biologically productive). Any wood will do, although birch is especially good, but avoid trees which are allelopathic (give off chemicals to prevent other plants growing near them) such as walnut. If possible, lay the line north to south to allow planting to benefit from the full rotation of the sun, or plant different plants on each side to suit their preference for more sun or shade.

Add smaller logs on top of the large ones, then add a layer of sticks and twigs, followed by leaf litter or compost. Finally, replace the topsoil [3–4].

Cover in mulch (straw, composted bark or home-made compost) and allow to sit through a few days of rain to settle. Then plant into it. [5]

In a warm climate it makes sense to sow seeds and cover them with straw just before it rains, and in cooler climates you should wait until the air temperature is right for your seeds, and if you plant by the moon, sow them just after the new moon.

1
2
3

4

5

Project: Start an integrated garden

Sometimes called a forest garden because the long-term relationships that are established between plants are similar to those found in a forest, but in no way needing an actual army of trees, this approach is a based on permaculture, which is a way of farming intended to be permanent, sustainable and self-sufficient. This is a long-term project so please don't feel the need to do it all in a weekend. You might plant the trees first and protect the soil below with ground-cover plants, and then pause. You can come back and plant the other layers over weeks, months or seasons, or even as you propagate your choices from seed or cuttings.

This approach translates well from agriculture to city gardens where space and light are at a premium because it relies on building up vertical layers and using mainly perennials, shrubs and trees, with occasional annuals brought into sunny edges.

---

1. Soil: Start by preparing the ground with a layer of cardboard topped with compost (*see page 138*), and then add an extra 5cm/2in depth of manure or organic matter, if your soil has not been looked after in recent years. Plant up with ground-cover plants (see below), which grow quickly and improve the soil as their root systems develop. They can be cut down after flowering and their whole biomass of leaves, stems and flowers left to rot down in situ. This helps to build the soil and encourages beneficial bacteria. If you have twigs and sticks, leave those on the ground, too, to help encourage beneficial fungi, or you can buy mycorrhizal fungi in powder form and add a little to the planting pits of the shrubs and trees.

2. Ground cover: This layer includes low-growing plants that are known as ground cover because they protect the soil and act like a living mulch. They include plants such as wild strawberry (*Fragaria vesca*), comfrey (*Symphytum officinale*), borage (*Borago officinalis*) and sweet violets (*Viola odorata*) and are key to the success of the forest garden, while also forming an essential part of the food chain by attracting beneficial insects. As opportune plants invite themselves in, you can decide if they should stay. Some, like rocket or nasturtium, will be self-seeders from the veg patch, while others will come in through bird droppings. Many so-called weeds, such as ivy (*Hedera*), ground elder (*Aegopodium podagraria*), nettles (*Urtica*), rosebay willowherb (*Chamaenerion angustifolium*), good King Henry (*Chenopodium bonus-henricus*) and lemon balm (*Melissa officinalis*) make great foraged foods and herbal remedies too. It is essential, though, to know your plants before eating any part of them.

3. Trees: In a large garden you can plant trees that will grow as tall as 10m/33ft and as wide as 6m/20ft, but in a smaller space you could begin with tall shrubs to create the canopy (see step 4, overleaf). Choose a multifunctional blossoming fruit or nut tree with a canopy that will allow light through to the layers below in summer. Cherries, apples, pears and plums all fit the bill, or try a false

acacia (*Robinia*), which will also fix nitrogen. Make sure you leave enough space for the trees to grow to maturity and still allow light through by checking their final heights and spreads. You can also plant a few extra shrubs between them, as long as you leave plenty of space for each one to fill out.

4. Shrubs: Choose shrubs that you like to look at and with fruit or nuts to eat, which will attract wildlife and have other uses. The strawberry tree (*Arbutus unedo*), roses, blueberries (on acid soil), witch hazel (*Hamamelis*), elderberries (*Sambucus*), currants and gooseberries are all good candidates, or plant hazel (*Corylus avellana*) for its nuts and stems to coppice for plant supports, stakes or even firewood. Choose smaller shade-tolerant shrubs such as *Gaultheria shallon*, *Sambucus racemosa*, the guelder rose (*Viburnum opulus*), *Fuchsia magellanica* or goji berries to plug gaps, and cordoned fruit trees or edible hedge plants to grow along the boundaries.

5. Herbaceous perennials: The herbaceous layer is a fast-growing, hungry combination of perennials and bulbs, so apply a deep mulch of compost around them as you plant, to feed the soil and get things going. Over time, the plants will create their own nutrients as their top growth falls on to the soil in autumn and composts down. Choose your herbaceous layer for beauty and usefulness. Place sun-loving plants or bulbs in the sunniest spots and those that need more water under the 'drip line' at the edge of the tree and shrub canopies – this is where water drips when it rains from the taller plants on to the ground. Your choices could include edible plants, such as *Agastache foeniculum*, *Aquilegia vulgaris*, *Angelica sylvestris*, *Malva sylvestris* or the sedum *Hylotelephium spectabile*. You can also include species that are good for bees and tolerate being walked on occasionally, including *Alchemilla mollis*, *Geranium macrorrhizum* and *Tiarella cordifolia*. Some, like nettles, can be used for dyes and cordage, or you could include soapwort (*Saponaria officinalis*) and philadelphus to make soap.

6. Climbers: Plant climbers such as hops, runner beans or passion fruits to scramble up medium-height trees to the sunlight. As the canopy matures, shade-lovers, including honeysuckles (*Lonicera*) and clematis, will continue to thrive and support wildlife. Plant these about 60cm/24in from the base of the trunk to allow the roots to gain access to water and nutrients, and tie the young climbers on to twiggy sticks or a cane leant at an angle against the tree to help them reach the lower branches. You can also plant climbers against a fence or wall, or allow them to scramble through shrubs as they would in the wild. Just cut them back if they become too vigorous to prevent them swamping their host.

7. Roots and fungi: Roots, including potatoes, onions, ginger and horseradish, can thrive on the floor of your integrated garden. Simply clear an area in semi-shade and plant them in staggered rows to maximize use of the space. You can add a further layer using fungi. Fungi will naturally gather where conditions are moist and dark and some will be edible, but others will be poisonous, so do check before eating any. For a delicious addition to the garden, you can buy edible mushrooms such as oyster mushrooms online: these come as a simple kit and are inoculated into fallen branches or logs. Fungi have amazing properties, and some say they may even be able to clean up plastic pollution in the oceans. There are some excellent books on them if you are interested in their applications, but in the meantime, they are a worthy addition to any sustainable garden.

Gardening in tune with the rhythms of the earth creates a great feeling of connection.

Garden in tune with the seasons and the moon

As gardeners we are much more aware of the seasons and the weather – we get used to the rhythms of the year and begin to understand the reasons behind harvest gluts in late summer and the 'hungry gap' in early summer, before many vegetables are ready to pick. We also notice the effects of the weather on our gardens, be it a hot dry period in summer, a late spring when we are waiting for the first buds to open, or a very wet winter when mature trees with weakened saturated roots can be toppled in a storm.

Through experience, many of us are attuned to the natural order of plant growth: flowering may be late in a cold dry year, for example, but snowdrops will always flower before daffodils, and both will bloom before alliums, as a rule. In the same way, we may be aware of the effect of the moon on the tides, as the full moon 'pulls' the water up to make bigger swells, and if you practise yoga you may also be aware of the moon's effect on your body, which is at least 60 per cent water. This gravitational power of the moon is also used in biodynamic gardening, where you plant and sow according to the moon phases. The method follows the moon's rhythms to create a gardening calendar, with days marked as beneficial to sow seeds when the moon is waxing and the light is increasing on the way to a full moon, and days to plant perennials and roots when it is waning from a full to a new moon. While science has not yet proven the effects, biodynamic gardeners and an increasing number of biodynamic wine producers swear by the results. If you are interested in working with the natural power of the moon to help your plants thrive, look up your local biodynamic association for more information.

Gardening for your gut

In chapter one, I explain how looking after your soil is like giving your garden a healthy gut, and that our bodies are full of different fungi and other microbes that make up our microbiota. In fact, we are learning that the body mass we carry around is made up of just 43 per cent human cells and the rest comprises microbes, which keep us healthy. We are also understanding more and more about the importance of our gut health and the organic food movement is a response to the understandable desire to avoid poisons in our diet.

In our own gardens we can take this to the next stage and combine what we know about healthy guts with healthy gardens to grow an all-round healthy mini biome. As children, we may have been told to eat our carrots to see in the dark, or to eat spinach to make us strong, and this old wisdom lives on while the science catches up and proves it to be true. These days, we are advised to eat a 'rainbow' by including in our diets a diversity of plants in an array of colours to ensure we get a good spread of vitamins and minerals. However, the science behind this also shows that we need more than vitamins and minerals to stave off diseases such as cancer, dementia and mental illness. Our immune system begins in our gut and relies on the healthy balance of our own microbiotic ecosystem.

When deciding what to grow, we can go a step further and not only select what we like to eat but add to the list food that our bodies, and our microbes, would like to be fed, too. We can ensure that we are as healthy as we can be with enough exercise and rest, sunlight, food, and love, of course, plus just enough stress to be in peak fitness. And we can do the same for plants by learning what food they need to be healthy, how they perform under different conditions and how we can reproduce those conditions to grow the plants that provide us with the most nutrients.

Studies show that plants produce nutrients and phytochemicals (chemical compounds of benefit to plants and humans) that our own microbiota also need when they are at key stages in their development and when they are under just enough stress from their surroundings or from insect browsing. The science behind this is beautiful although perhaps beyond the scope of this book, but the evidence boils down to not over-fertilizing your plants and allowing some insect attacks to stimulate natural defence responses that trigger the release of phytochemicals. The perfect combination of good intention and benign neglect that I know plants get in my garden!

Grow food that your gut will thank you for in conditions that produce the most nutrients.

Our hedgerows are full of leaves and berries to forage.

A simple nettle tea is a great cleanser.

Foraging food for free

Foraging makes sense on so many levels. Not only can we help ourselves to nature's bounty with very little effort, but by grazing the land we are tending it and becoming an integral part of it, eating what is available locally and paying attention to what the land is offering us. Eating a wide variety of plants also feeds our gut microbiota and increases the number and diversity of beneficial bacteria, which in turn makes us healthier, so it makes sense to add a little of many different foods to our plates.

In spring, the tender new growth of nettles and cleavers are great cleansers for the liver and sources of key vitamins; the new leaves of hawthorn are fresh and tender; and the foliage of cow parsley and sorrel can be added to salads or risottos.

We can even combine foraging with weeding and eat the wild edible plants that clutter tidy borders. The young leaves of ground elder are delicious in omelettes and are very cleansing for the gut – the Romans originally exported this plant across the sea to combat gout caused by excessive banqueting. Fresh chickweed and ground ivy (*Glechoma hederacea*) leaves are other tasty examples, while wild violets are pretty in salads or drinks, and lemon balm (*Melissa officinalis*) makes a refreshing tea and is said to be a cure for anxiety.

In the autumn, there are many delicious types of mushroom to forage, or you can grow your own using simple kits. I hope that other countries will copy France in the future, where you can take your foraged fungus to the pharmacist to be identified and to ensure that it's not poisonous. In the meantime, with all foraging, do make sure you can identify your species and their lookalikes, since many mushrooms and plants are powerful medicines or poisons, and mistakes can be fatal.

Rosehips, sloes and blackberries provide foraged autumn vitamins.

Regenerating vegetables

It is possible to have your celery and eat it! Too often we throw out the ends of our vegetables when we could add them to a stock to make nutritious soups, stews or risottos, or recycle them to grow more food for free. So, before you pile up the waste to put in the compost or bokashi bin, take a look at what's left of your veggies and sort out those that you can use.

How often have you discarded your onions or garlic because they were sprouting green shoots or your potatoes because they were growing eyes and turning a gentle shade of green? Onions and garlic are bulbs, and potatoes are tubers that will grow if you put them into soil (or even the compost bin). You can also experiment by inserting the tops of beetroots and carrots into soil to grow beet leaves and carrot tops, both of which are highly nutritious.

The benefits of reusing organic vegetables are that they are pesticide-free and contain higher levels of health-promoting phytochemicals (*see page 148*), but even many everyday supermarket vegetables will regenerate, given the right conditions.

In spring, when you chop off the end of your celery, put the base in water, then wait until little white roots appear. Pot it up into some compost, set it in a fairly cool place in the garden and keep it moist. After a few weeks

new shoots will start to grow, which you can harvest immediately to add to salads and soups, or pick from the side like a cut-and-come-again salad to encourage more new growth. You can also cut the bottom of your lettuce off in the same way, stick it in some compost, and watch it reshoot. You may soon find you have enough to swap with friends.

Supermarket herbs are another great source of regenerated food. Basil is often over-planted, with lots of stems fighting it out in tiny pots. It's easy to gently tease apart clusters of stems and their roots and divide them up into several pots to grow on into lovely, lush plants that can be harvested through the season. Or take cuttings from your supermarket mint by cutting a stem a few leaves down from the top, just below a leaf joint. Strip off the bottom two leaves to prevent them from rotting and pop the stem in a glass of water. It will soon grow long white roots, and you can then transfer it to a pot of compost.

Tomatoes and peppers contain lots of seeds, which you can save from your favourite varieties. Simply rinse them to remove the jelly-like coating and dry them on a piece of paper towel. They will keep in an envelope in a cool dry frost-free place over winter.

Celery, lettuce, mint and tomatoes are all easy to regenerate.

Conclusion

From the thrill of a seed germinating or the harvesting of a fruit, to the frustration of a plant failing and the desire to try again, our innate gardener's curiosity keeps us fascinated by the natural world – its workings and its gifts. To hold this book in your hand shows you share that curiosity and are on your way to a lifetime of learning and wonder.

I hope the ideas, projects and images in these pages have helped to inspire you to create beautiful spaces and to navigate some of the choices we face every day when we garden, from what and where to plant, to what and how to build. Our seemingly small everyday choices combine to make a disproportionate difference, and in opting for beauty and sustainability over convenience and expedience, we are improving the health of ourselves and our environment. Nature is inherently wonderful and infinitely powerful, and once we understand and experience how to play our part, so are we.

The path to sustainable gardening is a lifelong journey.

Index

Page numbers in *italics* indicate an illustration; those in **bold** indicate a main section of project.

A

acacia, false *see Robinia pseudoacacia*
Aesculus hippocastanum 113
Agastache foeniculum 144
air pollution, helping 40
Alchemilla mollis **118–19**, 144
alder, Italian *see Alnus cordata*
algal blooms 114
Allium **118–19**
Alnus cordata (Italian alder) 114
aluminium 47
Amelanchier 113
Ammi majus **118–19**
Angelica sylvestris 144
Anthriscus sylvestris (cow parsley) 117, 150
Aquilegia vulgaris 144
Arbutus unedo (strawberry tree) 113, 144
Aronia arbutifolia (red chokeberry) 114
aspect of garden *14*, 15

B

badger problems 43
Baptisa australis 117
basil 33, 153, *153*
bathtubs 91, **92–5**
bay *see Laurus nobilis*
bean family 114
beans 136, 144
  supports *137*
bees 40, *112*
beetroot 152
benches 34, *41*, *49*, *82*
berries 30, *145*
besom broom **80–1**
Betula (birch) 114
bird boxes **102–3**
bird feeders 75
birds, encouraging 17, 30
black locust *see Robinia pseudoacacia*
blackberries *151*
blackthorn 53
bluebells *10*
blueberries 144
Borago officinalis (borage) 142
boundaries **38–43**, **44–55**
boxes, reusing 91, **96–7**
bricks 44, *50*, *51*
buckwheat as green manure 60
buddleia (butterfly bush) 114, 117
bug hotel *48*
bulbs 131
bulrush *see Typha*
butterfly bush *see* buddleia
butts 17, *69*, *69–70*, 77

C

cabbages 136, 138
calabrese 135
Calamagrostis x *acutiflora* 'Karl Foerster' **118–19**
calendar for gardening 147
canna lily 70
carbon
  footprint of cement 47
  footprint of lawnmowers 20–1
  footprint of mortar 40
  peat locking up 59
  plants turning sunlight into 20
  soil locking up 56, 58
  trees locking up 48, 111
cardboard tubes *78*, 78
Carex spp. 15
carrots *134*, 134, 152
celery 152–3, *153*
Chamaenerion angustifolium (rosebay willowherb) 142
chickweed 150
child labour 85
chillies 92–3
chives 33, 138
chokeberry, red *see Aronia arbutifolia*
cleavers 150
Clematis 114, 144
  C. 'Alba Luxurians' 30
climbers 30, 107, 114–15, 144
cloches *74*, 75
clover as green manure 60
coffee granules 60, *61*
coir pots 77
cold frame 78, 79
comfrey *see Symphytum officinalis*
comfrey tea **62–3**
compost 17, 20, 58, 59, *59*, 114
compost bags 75
compost bin for leaves 64–5
concrete 51
containers **30**, *31*, *46*, 91, *91–2*
  recycled **92–3**
Corylus avellana (hazel) 53, 112, 144
courgettes 136
cow parsley *see Anthriscus sylvestris*
crab apples *see Malus*
crinkle-crankle wall 40
Crocosmia **118–19**
currants 144
cut-and-come-again leaves 134
cuttings **124–5**

D

decking 47
diet for health 148
dining area 37
ditches 71
drainage 36
drainpipes 91
drought-tolerant plants *16*
dry soil 15

E

egg shells 60, *60*
Elaeagnus angustifolia (oleaster) 114
elder *see Sambucus*
electric tools 21
Ethical Trading Initiative 85
Eucryphia glutinosa 113
Euphorbia **118–19**

F

F1 hybrids 126
fedges 43
fences 40–3, *41*
fertilizers 114
filter beds 70
food waste, regenerating 152
foraging 150
'forest' garden 142–3
fossil fuels 20
foxglove 117
Fragaria vesca (wild strawberry) 142
frame for succulents **98–9**
freecycling 88

fruit trees *37*, 112–13, 142, *143*, 144
Fuchsia magellanica 144
fungi
growing for eating 144
in soil 58
furniture **86–91**

G
gabions *45*
garden room *26*
gardeners, employing 19
gardening calendar 19
garlic 152
gates *43*
Gaultheria shallon 144
Geranium **118–19**
G. macrorrhizum 144
see also Pelargonium
glass recycling *49*
glasshouse 20
plastic for *74*
vegetables for 135
Glechoma hederacea (ground ivy) 150
gloves, bamboo 24
goji berries 144
good King Henry 142
gooseberries 144
grasses, ornamental 117
gravel areas 35, *50*, **128–9**
Great Dixter 108
green manures 60
ground elder 142, 150
ground ivy *see Glechoma hederacea*
guelder rose *see Viburnum opulus*
gut health **148–53**

H
Hamamelis (witch hazel) 144
hawthorn 53, 150
hazel *see Corylus avellana*
heating glasshouses 20
Hedera (ivy) 107, 114, *115*, 142
hedges 40
'dead' **54–5**
edible **52–3**, 144, *150*
Helenium (sneezeweed) **118–19**
herbs 19, 30
making a spiral *32*, **33**
in paving 35
from supermarkets 153
in wine box **96–7**
hessian 78
Hippophae rhamnoides (sea buckthorn) 53, 114
honeysuckle *see Lonicera*
hops 144
hornbeam **118–19**
horticultural fleece 78
hose, leaky 70
houseleek *see Sempervivum*
hügelkultur mound **140–1**
hurdles 42
hydrangea, climbing 107
Hylotelephium spectabile (sedum) 144

I
integrated garden **142–4**
internet plant sourcing 108
ivy *see Hedera*

K
kale 138
Khadi fabric 89, *89*
kneeler 24
Kniphofia **118–19**
Koelreuteria paniculata 113

L
ladders as plant support 91
Lathyrus odoratus (sweet pea) 114
Laurus nobilis (bay) 30
lawn mower emissions 20–1, *21*
lawns 38
making a spiral **28**, *29*
raking *21*
layered planting *2*, 17, 58, 69, 114, *116*, 117, **120–1**
leaf blowers 21
leaves, composting **64–5**, 114
leeks *135*
legumes as green manure 60
lemon balm *see Melissa officinalis*
lettuce *153*, 153
lime *see Tilia*
Liriodendron tulipifera 113
'local' buying 47
London soil 15
Lonicera (honeysuckle) 114, *115*, 144
loosestrife *see Lythrum salicaria*
Lythrum salicaria (purple loosestrife) 117

M
Madoo Conservatory Garden 108
Malus (crab apple) 53, 112, *112*
M. domestica 113
Malva sylvestris 144
manure
green 60
compost 58
materials for boundaries **44–55**
meditation 27
Melissa officinalis (lemon balm) 142, 150
microbiota
in gut **148–53**
in soil 58
moon, gardening with phases 147
mortar 35, 40, *41*, *51*
moss 114
moss on walls 40
mound culture **140–1**
mulching 58, 144
mushroom compost 58
mushrooms 144, 150
mustard greens 136
mycorrhizal fungi 58, 111
Myrtus communis (myrtle) 30

N
nasturtium 136, 142
nettle tea 63, *150*
nettles 142, 144, 150
nitrogen fixation 60, 114
nitrogen run-off 114
no-dig methods 58, **138–9**
noise pollution from garden tools 21
north-facing aspect 15
nurseries 108, *109*

O
observing your garden 27
oleaster *see Elaeagnus angustifolia*
onions 152

P
packaging, plastic-free 77
paint 89–90
passion fruit 144
paths 38, *39*, *50*, 74, 155
pavers *34*
paving 35
materials for 45–7
permeable 69
pea family 114
peas 134, *134*
peat 59
Pelargonium 20, *31*
penknife 24
peppers 153
perennials **117–21**
splitting 122
permaculture 142
pesticides 114
pests 136, 138
petrol tools 21

pH of soil *58*, 58
phacelia as green manure 60
Philadelphus 144
phytochemicals 148, 152
planning the garden **14–17**, **36–7**
for enjoyment 18–19
planting **104–53**
where to put items and plants 36, 38
plant fairs 108, *108*
plant insulators 78
plant pots 76–7
planters *see containers*
plants
choosing 36
swaps 108
for water management 69
plastic
bottles, use for 75
in the garden 75
tools 24
polyester 87
polyurethane 87
pond liners 71
ponds *69*, 70–1, *71*, 91, **94–5**
potatoes 134, 152
Prunus
P. avium 113
P. serrula 115

R
rabbit problems 43
rain gardens **72–3**
rainwater 17
raised beds 136
rake 24
reduce/reuse/recycle 12, 47, 70, 78, **87–103**, 152
see also upcycling
Robinia pseudoacacia (black locust/ false acacia) 114, 142
rocket 142
rose hips 145, 151
rosebay willowherb *see Chamaenerion angustifolium*
rosemary 33
roses 114, 144

S
sage 33
Salix (willow) 15
dome **100–1**
fences 42
S. alba 113
Salvoweb 88
Sambucus (elder) 53, 144
Sanguisorba **118–19**
Sarcococca confusa (sweet box) 30
screening 36, *37*
sculpture, oak *45*
sea buckthorn *see Hippophae*
seating 27, *39*, *82*, *86*, *87*, *89*, 91
see also benches
sedges 15
sedums 30, 98–9, 144
seed blocks *60*
seed pots 78
seedlings 78
seeds 122, **126–7**
Sempervivum (houseleek) 30, 98–9
sewing machine bases 88
shady areas 36
shovel 24
shrubs 114–15, 144
sinks for water 72–3, *74*
sloes *151*
sneezeweed *see Helenium*
soapwort *see Saponaria officinalis*
soil **56–67**
conditions 15
feeding 59
health 58
identifying 58
in integrated fardens 142
solar heating 20
sorrel 150
south-facing aspect 15
spade 24
star jasmine see *Trachelospermum*
steel furniture 87
stone *51*
for paving 45, 47
table *82*
for walls 40, 42, 44, *84*
straw bales *91*, 91
strawberry, wild *see Fragaria vesca*
strawberry tree *see Arbutus unedo*
succulents **98–9**
sunny aspect 17, 36
swales 71
sweet box see *Sarcococca confuse*
sweet pea *see Lathyrus odoratus*
Symphytum officinale (comfrey) 142

T
table *82*, 88, *89*
terracotta pots *76*, 76
Tiarella cordifolia 144
tidiness habit 13
Tilia cordata (lime) 112, 113
tins for planters 91
tomato leaf tea 63
tomatoes 126, 134, 153, *153*
tools 21, **22–4**, *23*, *77*
Trachelospermum jasminoides (star jasmine) 30
tree bags 70
trees 15, 17, *110*, 110–13
for bees and wildlife 112–13
blossom calendar 113
decidous 17
for hedges 53
in integrated gardens 142–4
placing 38
for structure 18
for temperature control 111
for water management 69
trowel *24*
tunnel, growing *45*
Typha (bulrush) 70

U
underground 15
upcycling 88–9
upholstery fabrics 87, 89

V
vegetables 19, 132–7, *149*
Viburnum opulus (guelder rose) 144
Viola odorata (sweet violet) 142, 150
Vipots 77

W
'wabi sabi' 85
walls 40, *41*, 84
water **68–74**
collecting and reusing 69–70
preserving 69
recycling 70
for wildlife 70–1
water resources 17
water table 15
watering can 24, *76*
wicker 89
wildflowers 38, **130–1**
wildlife 36, 54, 70–1, 112–13, 114, 117
willow *see Salix*
windbreaks 40, 43
windy aspect 15, 36
wine box reuse **96–7**
witch hazel *see Hamamelis*
wood for boundaries 44, 48
wormery 60, *66*, 67

Y
yellow rattle 131

Further reading

If the holistic approach resonates with you, here are just a few books that will take you further.

Hedgerow Medicine: Harvest and Make Your Own Herbal Remedies, Julie Bruton-Seal and Matthew Seal (Merlin Unwin, 2008)

Wayside Medicine: Forgotten Plants and How to Use Them, Julie Bruton-Seal and Matthew Seal (Merlin Unwin, 2017)

The Sustainable Sites Handbook: A Complete Guide to the Principles, Strategies, and Best Practices for Sustainable Landscapes, Meg Calkins (Wiley, 2012)

Creating a Forest Garden: Working with Nature to Grow Edible Crops, Martin Crawford (Green Books, 2010)

How to Grow Perennial Vegetables: Low-maintenance, Low-impact Vegetable Gardening, Martin Crawford (Green Books, 2012)

Feng Shui: Secrets of Chinese Geomancy, Richard Creightmore (Wooden Books, 2011)

Charles Dowding's Vegetable Course, Charles Dowding (Frances Lincoln, 2012)

Naturalistic Planting Design The Essential Guide: How to Design High-Impact, Low-Input Gardens, Nigel Dunnett (Filbert Press, 2019)

The More Beautiful World Our Hearts Know is Possible, Charles Eisenstein (North Atlantic Books, 2013)

The Dry Gardening Handbook: Plants and Practices for a Changing Climate, Olivier Filippi (Filbert Press, 2019)

Thus Spoke the Plant: A Remarkable Journey of Groundbreaking Scientific Discoveries and Personal Encounters with Plants, Monica Gagliano (North Atlantic Books, 2018)

How To Read Water: Clues & Patterns from Puddles to the Sea, Tristan Gooley (Sceptre, 2016)

The Songs of Trees: The Stories from Nature's Great Connectors, David George Haskell (Viking, 2017)

Gaia's Garden: A Guide to Home-Scale Permiculture, Toby Hemenway (Chelsea Green Publishing, 2009)

Sowing Beauty: Designing Flowering Meadows from Seed, James Hitchmough (Timber Press, 2017)

Collins Tree Guide, Owen Johnson (HarperCollins, 2004)

Braiding Sweetgrass: Indigenous Wisdom, Scientific Knowledge and the Teachings of Plants, Robin Wall Kimmerer (Penguin, 2020)

Construction for Landscape Architecture, Robert Holden and Jamie Liversedge (Laurence King, 2011)

The Well-Tempered Garden, Christopher Lloyd (Viking, 1985)

Novacene: The Coming Age of Hyperintelligence, James Lovelock (Allen Lane, 2019)

Concise British Flora in Colour, William Keble Martin (Ebury Press, 1965)

Natural Selection: A Year in the Garden, Dan Pearson (Guardian Faber, 2017)

The Environmental Design Pocketbook, Sofie Pelsmakers (RIBA Publishing, 2012)

Design Studio Vol. 1: Everything Needs to Change: Architecture and the Climate Emergency, Sofie Pelsmakers (RIBA Publishing, 2021)

The Overstory, Richard Powers (William Heinemann, 2018)

The Illustrated History of the Countryside, Oliver Rackham (Weidenfeld & Nicolson, 1994)

Planting in a Post-Wild World: Designing Plant Communities for Resilient Landscapes, Thomas Rainer and Claudia West (Timber Press, 2015)

Entangled Life: How Fungi Make Our Worlds, Change Our Minds, and Shape Our Futures, Merlin Sheldrake (Vintage, 2020)

Fantastic Fungi: How Mushrooms Can Heal, Shift Consciousness, and Save the Planet, edited by Paul Stamets (Earth Aware, 2019)

The Well Gardened Mind: Rediscovering Nature in the Modern World, Sue Stuart-Smith (William Collins, 2020)

Reclaiming the Wild Soul: How Earth's Landscapes Restore Us to Wholeness, Mary Reynolds Thompson (Write the Damn Book, 2019)

Wilding: The Return of Nature to a British Farm, Isabella Tree (Picador, 2018)

The Hidden Life of Trees: What They Feel, How They Communicate, Peter Wohlleben (Greystone Kids, 2016)

Green Building Handbook Vol. 1: A Guide to Building Products and their Impact on the Environment, Tom Woolley and Sam Kimmins (Routledge, 1997)

Websites

https://www.biodynamic.org.uk
https://www.biodynamics.com
https://www.permaculture.org.uk

First Published in 2022 by Frances Lincoln,
an imprint of The Quarto Group.
The Old Brewery, 6 Blundell Street,
London N7 9BH, United Kingdom.
T (0)20 7700 6700
www.Quarto.com

Commissioning editor: Helen Griffin
Design by Rachel Cross

A catalogue record for this book is available from the British Library.

ISBN 978-0-7112-6788-6

9 8 7 6 5 4 3 2 1

Typeset in Avenir

Printed and bound in China

About the author

Marian Boswall is a leading landscape architect and horticulturalist, was a lecturer in Historic Garden Conservation at Greenwich University for several years and is a co-founder of the Sustainable Landscape Foundation. She writes and lectures on sustainable design and was awarded the Garden Columnist of the Year in 2019. She has also been featured as a *Country Life* 'Top 50' Garden Designer, *House & Garden* 'Top 50' UK Garden Designer and *Country and Town House* 'Top 10' Garden Designer.

Acknowledgements

Huge thanks to the friends and clients that have let Jason and I use their gardens to photograph some of my favourite sustainable ideas. So many people have helped directly and indirectly from providing plants, to inspiration or support, thank you! Here are just a few:

Jason Ingram, Lucy Bellamy, the fabulous team at Frances Lincoln, Sasha Georgiev, Lady Nevill, Robyn and Mark Reeves, Charlotte and Donald Molesworth, Katie Platt-Higgins, Kate and Aidan Kong, Jim and Sheila Briggs, Martin Crawford, Matt Somerville, Eliza Pelham and Ed Conway, Richard Creightmore and Jewels Rocka, Abbey Physic Community Garden, Tillingham Biodynamic Winery, The Blackthorn Trust, Simon Sutcliffe, Elizabeth Cairns, Sarah Salway, Kate and Hugo Fenwick, Ange Snape, Imogen McAndrew and the team at MBLA, Adam White, Arit Anderson Andrée Davies and the SLF Team, and my family, who thrive like my plants, with minimal intervention.

Profits from the sale of this book will go to the Sustainable Landscape Foundation.